Edexcel AS | UNIT 1

Physics

Physics on the Go

Mike Benn

Philip Allan Updates, an imprint of Hodder Education, an Hachette UK company, Market Place, Deddington, Oxfordshire OX15 0SE

Orders

Bookpoint Ltd, 130 Milton Park, Abingdon, Oxfordshire OX14 4SB
tel: 01235 827720
fax: 01235 400454
e-mail: uk.orders@bookpoint.co.uk
Lines are open 9.00 a.m.–5.00 p.m., Monday to Saturday, with a 24-hour message answering service. You can also order through the Philip Allan Updates website: www.philipallan.co.uk

© Philip Allan Updates 2009

ISBN 978-0-340-94826-2

First printed 2009
Impression number 5 4 3 2 1
Year 2014 2013 2012 2011 2010 2009

This guide has been written specifically to support students preparing for the Edexcel AS Physics Unit 1 examination. The content has been neither approved nor endorsed by Edexcel and remains the sole responsibility of the author.

Typeset by Pantek Arts Ltd, Maidstone, Kent.
Printed by MPG Books, Bodmin

Hachette UK's policy is to use papers that are natural, renewable and recyclable products and made from wood grown in sustainable forests. The logging and manufacturing processes are expected to conform to the environmental regulations of the country of origin.

Contents

Introduction

About this guide 4
The specification 4
The unit test 5
Revision 6

■ ■ ■

Content Guidance

About this section 8
Mechanics...9
   Rectilinear motion 9
   Scalar and vector quantities...16
   Forces..20
   Work, energy and power 29
Materials...33
   Properties of fluids...33
   Tensile behaviour of solid materials 42

■ ■ ■

Questions and Answers

About this section...52
Test Paper 1...53
Test Paper 2...67

Introduction

About this guide

This guide is one of a series covering the Edexcel specification for AS and A2 physics. It offers advice for the effective study of Unit 1: Physics on the Go. Its aim is to help you *understand* the physics — it is not intended as a shopping list, enabling you to cram for the examination. The guide has three sections:

- **Introduction** — this gives brief guidance on approaches and techniques to ensure you answer the examination questions in the best way that you can.
- **Content Guidance** — this section is not intended to be a detailed textbook. It offers guidance on the main areas of the content of Unit 1, with an emphasis on worked examples. These examples illustrate the type of question that you are likely to come across in the examination.
- **Questions and Answers** — this comprises two sample unit tests. Answers are provided and, in some cases, distinction is made between responses that might have been given by an A-grade candidate and those of a typical grade-C candidate. Common errors made by candidates are also highlighted so that you, hopefully, do not make the same mistakes.

The effective understanding of physics requires time and effort. No one suggests it is an easy subject, but even those who find it difficult can overcome their problems by the proper investment of time.

The development of an understanding of physics can only evolve with experience, which means time spent thinking about physics, working with it and solving problems. This book provides you with the platform to do this. If you try all the worked examples and the unit tests before looking at the answers (no cheating!), you will begin to think for yourself as well as develop the necessary technique for answering examination questions. In addition, you will need to learn the basic formulae, definitions and experiments. Thus prepared, you will be able to approach the examination with confidence.

The specification

The specification states the physics that will be examined in the unit tests and describes the format of those tests. This is not necessarily the same as what teachers might choose to teach (or what you might choose to learn).

The purpose of this book is to help you with Unit Test 1, but don't forget that what you are doing is learning *physics*. The specification can be obtained from Edexcel, either as a printed document or from the web at **www.edexcel.com**.

The unit test

Unit Test 1 is a written paper of duration 1 hour and 20 minutes and carries a total of 80 marks. There are 10 objective (multiple-choice) questions, each worth a single mark, and a further 10 or 11 short and long questions worth between 3 and 12 marks. The test is designed to cover all the elements in the specification for the unit, and all the questions must be attempted. It is therefore important that you revise all sections before the examination.

The test will incorporate the assessment of 'How Science Works' — an exploration of how scientific knowledge is developed, validated and communicated by the scientific community (assessment objective AO3). It will also examine assessment objectives AO1 (knowledge and understanding) and AO2 (application of knowledge and understanding) in equal measure of about 30–40% of the marks.

A formulae sheet is provided with the test. Copies may be downloaded from the Edexcel website, or can be found at the end of past papers.

Command terms

Examiners use certain words that require you to respond in a particular way. You must be able to distinguish between these terms and understand exactly what each requires you to do. Some frequently used commands are shown below.

- **State** — a brief sentence giving the essential facts; no explanation is required (nor should you give one).
- **Define** — you can use a *word equation*; if you use *symbols*, you must state what each symbol represents.
- **List** — simply a series of words or terms, with no need to write sentences.
- **Outline** — a logical series of bullet points or phrases will suffice.
- **Describe** — for an experiment, a diagram is essential, then give the main points concisely (bullet points can be used).
- **Draw** — diagrams should be drawn in section, neatly and *fully labelled* with all measurements clearly shown, but don't waste time — remember it is not an art exam.
- **Sketch** — usually a graph, but graph paper is not necessary, although a grid is sometimes provided; axes must be labelled, including a scale if numerical data are given, the origin should be shown if appropriate, and the general shape of the expected line should be drawn.
- **Explain** — use correct physics terminology and principles; the depth of your answer should reflect the number of marks available.
- **Show that** — usually a value is given so that you can proceed with the next part; you should show all your working and give your answer to more significant figures than the value given (to prove that you have actually done the calculation).

- **Calculate** — show all your working and give *units* at every stage; the number of significant figures in your answer should reflect the given data, but you should keep each stage in your calculator to prevent excessive rounding.
- **Determine** — means you will probably have to extract some data, often from a graph, in order to perform a calculation.
- **Estimate** — a calculation in which you have to make a sensible assumption, possibly about the value of one of the quantities — think, does this give a reasonable answer?
- **Suggest** — there is often no single correct answer; credit is given for sensible reasoning based on correct physics.
- **Discuss** — you need to sustain an argument, giving evidence for and against, based on your knowledge of physics and possibly using appropriate data to justify your answer.

You should pay particular attention to diagrams, sketching graphs and calculations. Candidates often lose marks by failing to label diagrams properly, by not giving essential numerical data on sketch graphs and, in calculations, by not showing all the working or by omitting the units.

Revision

The purpose of this introduction is not to provide you with a detailed guide to revision techniques — there are many books written on study skills if you feel you need more help in preparing for examinations. There are, however, some points worth mentioning that will help you when revising for your physics AS exam. You should:

- familiarise yourself with what you need to know by asking your teacher and looking through the specification
- make sure you have a good set of notes — you cannot revise properly from a textbook
- learn all the equations indicated in the specification and be familiar with the equations that are provided at the end of each question paper, so that you can find them quickly and use them correctly
- make sure you learn definitions thoroughly and in detail, e.g. work is the product of force times distance *in the direction of the line of action of the force.*
- be able to describe (with a diagram) the basic experiments referred to in the Edexcel specification
- make revision active by writing out equations and definitions, drawing diagrams, describing experiments and performing lots of calculations.

Content
Guidance

This section is a guide to the content of **Unit 1: Physics on the Go**. It does not constitute a textbook for Unit 1 material. The unit is split into two topics — **mechanics** and **materials**. The main areas are:

Mechanics

- Rectilinear motion — equations of motion for uniformly accelerating bodies, displacement–time and velocity–time graphs, motion of objects moving freely under gravity, projectile motion
- Scalar and vector quantities — addition and resolution of vectors by scale drawing and mathematical methods
- Forces — the nature of forces, free-body force diagrams; Newton's first law of motion, equilibrium and centre of gravity; Newton's second law of motion for a body of fixed mass, gravitational field strength and weight; Newton's third law, Newton's third law pairs
- Work, energy and power — concept of mechanical work; relationship between work and energy, power, efficiency

Materials

- Properties of fluids — density, upthrust and viscous drag; laminar and turbulent flow; viscosity and Stokes's law
- Tensile behaviour of solid materials — stress, strain, Young modulus, force–extension (or compression) graphs, stress–strain graphs, elastic strain energy, properties of solids (brittle, ductile, hard, malleable, stiff and tough)

Quantity algebra

In the worked examples and in the answers to examination questions, quantity algebra is used throughout. This involves putting in the units of each quantity in an equation. For example, to calculate the density of a steel sphere of mass 17 g and diameter 16 mm we would write:

$$\rho = \frac{m}{V} = \frac{17 \times 10^{-3}\,\text{kg}}{\frac{4}{3}\pi(8 \times 10^{-3}\,\text{m})^3} = 780\ \text{kg m}^{-3}$$

Although Edexcel does not require you to use quantity algebra, you are strongly advised to do so because of the following advantages:

- It acts as a reminder to substitute consistent units — in this example 17 g is written as 17×10^{-3} kg and 8 mm as 8×10^{-3} m.
- It allows you to check that the units of the answer are correct — in this example kg in the numerator and m^3 in the denominator lead to the units of density, kg m^{-3}.

Mechanics

Rectilinear motion

This section covers the motion of objects in a given direction when they are either moving with a constant speed in that direction (i.e. constant velocity), or accelerating at a constant rate.

Definitions

- **Average velocity** is displacement divided by time.
- **Acceleration** is the rate of change in velocity.

Equations of motion (*suvat* equations)

The relevant quantities needed to represent the motion of an object in one dimension are:

- **displacement** s m
- **initial velocity** u m s^{-1}
- **final velocity** v m s^{-1}
- **acceleration** a m s^{-2}
- **time** t s

Using these symbols the definitions become:

- **Average velocity** $= \dfrac{(u + v)}{2} = \dfrac{s}{t}$

- **Acceleration** $= \dfrac{(\text{change in velocity})}{\text{time}} = \dfrac{v - u}{t}$

Combining the above definitions leads to a set of equations where, if any three of the quantities is known, the other two may be calculated. These are sometimes referred to as the 'suvat' equations and they are reproduced at the end of each Unit 1 examination in the formulae section.

- $v = u + at$
- $s = ut + \frac{1}{2}at^2$
- $v^2 = u^2 + 2as$

Worked example

An athlete starts from rest and accelerates uniformly for 4.0 s when she has reached a velocity of 8.0 m s^{-1}. Calculate:

(a) her acceleration

(b) the distance she has travelled during this time

> **Tip** When using the 'suvat' equations it is useful to write down all of the quantities with the given values, in this case:
>
> $u = 0$ m s^{-1} $v = 8.0$ m s^{-1} $t = 4.0$ s a and s are to be found.

Answer

(a) Use $v = u + at$

   8.0 m s^{-1} $= 0$ m s^{-1} $+ a \times 4.0$ s $a = 2.0$ m s^{-2}

(b) Use $s = ut + \frac{1}{2}at^2 = 0$ m s^{-1} $\times 4.0$ s $+ \frac{1}{2} \times 2.0$ m s^{-2} $\times (4.0$ s$)^2 = 16$ m

Acceleration due to gravity

In the late sixteenth century, Galileo Galilei performed his famous experiment, dropping a variety of different sized balls from the Leaning Tower of Pisa. The balls, when released simultaneously, hit the ground at the same time, showing that the acceleration due to gravity is independent of the mass of free-falling objects. Later experiments showed that a coin and a feather in an evacuated glass tube fell at the same rate, and the *Apollo 15* astronaut David Scott performed a similar test on the Moon using a hammer and a feather.

The value of the acceleration due to gravity has been accurately measured on the Earth's surface and, although there are small variations around the globe, the accepted value is usually given as 9.81 m s^{-2}. For the purposes of this guide all calculations use 9.8 m s^{-2} and if estimates are needed 10 m s^{-2} is adequate. This acceleration is usually referred to as 'little g'.

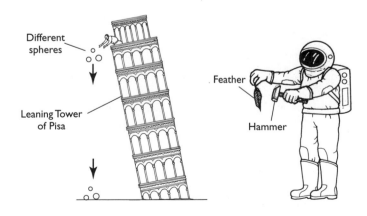

content guidance

The value of g can be found by timing an object falling freely from rest through a measured distance, and then using the appropriate equation of motion. In practice, because the distances, and hence the times, are usually relatively small, sensitive timing devices are required. A simple arrangement is shown in the diagram below where a switch de-energises an electromagnet and starts an electronic timer simultaneously. A steel sphere is released from the magnet and falls onto a 'trap-door switch' that opens to switch off the timer.

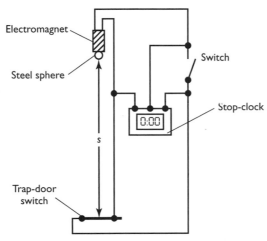

The equation $s = ut + \frac{1}{2}at^2$ is applied with s representing the distance fallen by the sphere. As the ball falls from rest, u equals zero and so the acceleration is found by rearranging the equation to give:

$$g = \frac{2s}{t^2}$$

There are several variations of this experiment. Using light gates with a computer interface, strobe photography and video-frame analysis are examples, but in all cases g is calculated from the time taken for a free-falling object to move through a measured distance.

Worked example

A stone is dropped into a well. The splash as it enters the water is timed as 1.8 s after release. Calculate:

(a) the depth of the well

(b) the velocity of the stone when it strikes the surface of the water

Answer

$u = 0 \text{ m s}^{-1}$ $a = g = 9.8 \text{ m s}^{-2}$ $t = 1.8 \text{ s}$

s (depth) and v are to be found.

(a) $s = ut + \frac{1}{2}at^2 = 0 + \frac{1}{2} \times 9.8 \text{ m s}^{-2} \times (1.8 \text{ s})^2 = 16 \text{ m}$

(b) $v = u + at = 0 \text{ m s}^{-1} + 9.8 \text{ m s}^{-2} \times 1.8 \text{ s} = 18 \text{ m s}^{-1}$

Horizontal and vertical motion

When a sky-diver jumps from a moving aeroplane, he will start to accelerate downwards but will also be moving horizontally at the speed of the aircraft. To explain the trajectory of the diver it is possible to consider the vertical and horizontal motions independently.

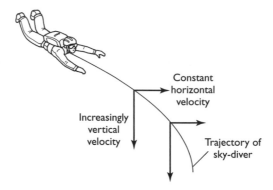

If air resistance is ignored, the sky-diver will accelerate downwards with a velocity of 9.8 m s^{-2} while at the same time continuing to move horizontally at the speed at which he left the plane. The above diagram illustrates the subsequent parabolic motion. It is possible to calculate the horizontal distance covered by a falling object by finding the time in free fall by applying the equations of motion in the vertical plane, and calculating the horizontal distance covered in this time when travelling at constant speed.

Worked example

A sea-eagle, carrying a fish in its talons, is flying horizontally with a velocity of 5.0 m s^{-1} at a height of 12 m above the sea. The fish wriggles free and falls back into the water. Calculate:

(a) the time taken for the fish to reach the sea

(b) the horizontal displacement of the fish during this time

Answer

(a) For the vertical motion: · $u = 0$ m s^{-1} $a = 9.8$ m s^{-2} $s = 12$ m

$s = ut + \frac{1}{2}at^2$ 12 m $= 0 + \frac{1}{2} \times 9.8$ m s$^{-2} \times t^2$ $t = 1.6$ s

(b) For the horizontal motion: $s = ut = 5.0$ m s$^{-1} \times 1.6$ s $= 8.0$ m

Displacement–time and velocity–time graphs

It is often useful to represent the motion of objects by plotting graphs of their displacement or velocities against time. In addition to illustrating the nature of the motion it is possible to calculate the values of displacement, velocity and acceleration from the graphs.

Displacement–time graphs

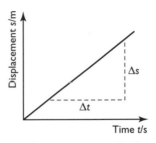

Graph for constant velocity

The diagram represents the displacement–time graph for an object moving at constant velocity.

$$\text{Velocity} = \frac{\text{displacement}}{\text{time}} = \frac{\Delta s}{\Delta t} = \textbf{gradient of the line}$$

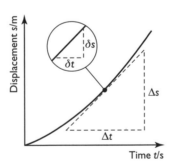

Instantaneous velocity for accelerating object

The gradient of the line is increasing with time. This means that the velocity is increasing and so the object is accelerating. The value of the velocity at any instant can be found by measuring the gradient of a tangent drawn at that instant.

$$\text{Instantaneous velocity} = \frac{\delta s}{\delta t} = \text{gradient of tangent} = \frac{\Delta s}{\Delta t}$$

Velocity–time graphs

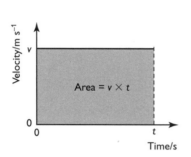
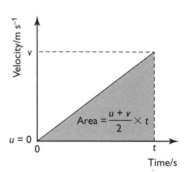

The graphs represent an object moving at constant speed and another with uniform acceleration.

$$\text{Acceleration} = \frac{\text{change in velocity}}{\text{time}} = \textbf{gradient of the line}$$

The first line has zero gradient, so the velocity is constant i.e. the acceleration is zero.

The gradient of the second line is constant, so the line represents uniform acceleration. A curve like the one drawn earlier for displacement against time, with a line of increasing gradient, but on a velocity–time graph, will indicate an increase in acceleration.

Tip When analysing displacement–time and velocity–time graphs always look at how the gradient is changing.

In the section on the equations for uniform motion we saw that:

$$\text{Displacement} = \text{average velocity} \times \text{time} = \frac{u + v}{2} \times t$$

The graphs show that the area under the line of each graph represents the displacement of the object during that period. This is also true for non-uniform motion, when the displacement can be found by estimating the area under the curve.

Worked example

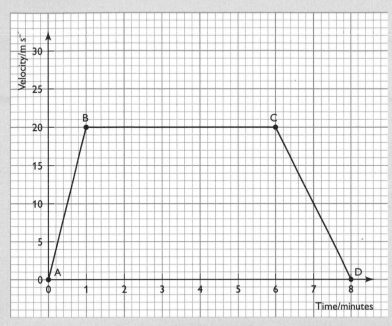

Velocity–time graph

The velocity–time graph represents the motion of a train as it travels from station A to station D. Describe the changes in the motion of the train, and calculate:

(a) the acceleration from A to B

(b) the acceleration from C to D

(c) the total displacement from A to D

Answer

The train accelerates uniformly from A to B, continues to travel at 20 m s⁻¹ until it reaches C, and then decelerates, uniformly, to D.

(a) acceleration = gradient = 20 m s⁻¹/60 s = 0.33 m s⁻²

(b) acceleration = gradient = –20 m s⁻¹/120 s = –0.17 m s⁻²

(The negative sign indicates that the acceleration is in the opposite direction to the velocity — i.e. it is decelerating.)

(c) displacement = area under the graph

= (½ × 20 m s⁻¹ × 60 s) + (20 m s⁻¹ × 300 s) + (½ × 20 m s⁻¹ × 120 s)

= 7800 m

Scalar and vector quantities

Definitions

- Scalar quantities have magnitude (size) only. Examples of scalars include *mass, the number of particles of a substance (moles), distance, speed* and *time.*
- Vector quantities have both magnitude and direction. Examples of vectors include *displacement, velocity, acceleration, fields* and *forces.*

The difference between scalar and vector quantities can be explained using the diagram below.

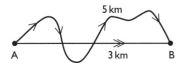

If you walk from A to B using the winding road, you will have travelled a distance of 5 km in no specific direction. This *distance* is therefore a scalar quantity.

On reaching B you will have been displaced from your starting position 3 km in the direction AB. This *displacement* is a vector quantity as it includes not only the size but also the direction.

If the journey took half an hour your average *speed* is the distance divided by the time; 5 km ÷ 0.5 h = 10 km h^{-1}, and as there is no implied direction it is a scalar quantity.

The average *velocity* is the displacement divided by the time; 3 km ÷ 0.5 h = 6 km h^{-1} **in the direction AB**, and therefore is a vector quantity.

It is important that the *direction* of a vector quantity is stated or clearly indicated on a diagram.

It is also important to be aware that in the equations for uniformly accelerated motion used earlier, the displacement, s, the initial and final velocities, u and v, and the acceleration, a, are all vector quantities, and time t is the only scalar. It is therefore necessary to indicate the direction of the vectors by allocating either a positive value (e.g. left to right or upward) and a negative value for quantities acting in the opposite direction (right to left or downward).

Worked example

A ball is thrown vertically upwards with an initial velocity of 8.0 m s^{-1}.
Calculate:
(a) the time taken for the ball to reach its maximum height
(b) the total upward displacement of the ball

Answer

Taking upward vectors as positive:

$u = +8.0$ m s^{-1}

$v = 0$ m s^{-1} (the ball stops momentarily at the top of its flight)

$a = g = -9.8$ m s^{-2} (negative as the acceleration is in a downward direction)

(a) $v = u + at$ 0 m s^{-1} = $+8.0$ m s^{-1} $-$ 9.8 m s^{-2} $\times t \rightarrow t = 0.82$ s

(b) $v^2 = u^2 + 2as$ (0 m s^{-1})^2 = $(+8.0$ m s^{-1})^2 $-$ 2 $\times$ 9.8 m s^{-2} $\times s$ $s = 3.3$ m

Scalar and vector addition

- Scalar quantities are added using normal arithmetic — for example, if you have 5 kg of potatoes in a basket and you add a further 2 kg of carrots, the total mass of the vegetables is 7 kg.
- To add vector quantities a diagram must be drawn showing both magnitude and direction of the quantities.

Worked example

An athlete runs due east for 4 kilometres and then due south for a further 3 kilometres. The run takes a total time of 20 minutes. Calculate:

(a) the distance travelled by the athlete

(b) her displacement from start to finish

(c) her average speed

(d) her average velocity

Answer

(a) Distance = 4 km + 3 km = 7 km

(b) The displacement can be found either by scale drawing, representing each leg of the run by a line where 1 cm, say, is equivalent to 1 km, and then measuring the resultant displacement; or, as the vector diagram includes a right-angle triangle, Pythagoras' theorem and basic trigonometry can be used.

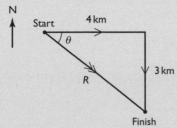

By measurement: $R = 5.0$ cm = 5 km; $\theta = 37°$

By calculation: $R^2 = (4$ km$)^2 = (3$ km$)^2$ $R = 5$ km; θ tan = ¾ $\rightarrow \theta = 37°$

The resultant displacement is 5 km in the direction E 37° S

(c) Average speed = 7 $\times 10^3$ m $\div$ 1200 s = 0.6 m s^{-1}

(d) Average velocity = 5 $\times 10^3$ m $\div$ 1200 s = 0.4 m s^{-1} in the direction E 37° S

Resolution of vectors

- A single vector can have the same effect as two other vectors that, when added together, would result in an identical vector.
- It is often convenient to represent a single vector as a pair of **component** vectors at right angles to each other.

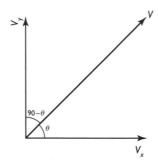

V is equivalent to the components V_x and V_y. Using trigonometry:

- $\cos \theta = V_x/V \rightarrow V_x = V \cos \theta$
- $\sin \theta = V_y/V \rightarrow V_y = V \sin \theta$

Tip If you feel unsure about when to use the sine or cosine component, you may like to always use the cosine of the angle between the vector and its components – so for the above example this could give $V_y = V \cos (90 - \theta)$ as an alternative to $V \sin \theta$.

Worked example

A gardener pulls a roller with a force of 200 N. The angle between the ground and the handle is 30°. Calculate the horizontal and vertical components of the force.

Answer

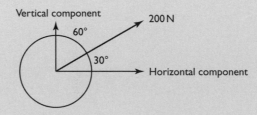

- Horizontal component = 200 N cos 30 = 173 N
- Vertical component = 200 N cos 60 (= 200 N sin 30) = 100 N

The roller behaves as if it is being pulled horizontally with a force of 173 N and being lifted up with a force of 100 N.

The resolution of a vector into a pair of components at right angles is a useful tool for many applications in physics. You will come across several applications later in this guide relating to the resolution of forces, and the A2 course requires the use of components of electric and magnetic fields.

You have seen that for bodies moving freely under gravity the horizontal and vertical motions can be treated independently. In the study of projectiles it is possible to gain information about the trajectories by taking horizontal and vertical components of the velocities.

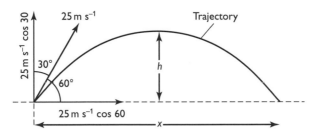

The initial velocity has components $u \cos \theta$ in the horizontal plane, and $u \cos (90 - \theta)$ (or $u \sin \theta$) in the vertical plane.

The vertical motion is always accelerating downwards at 9.8 m s^{-2} whereas the horizontal component of the velocity remains the same throughout the flight of the projectile.

Applying the equations of motion to the vertical components you can calculate the maximum height reached and the time of the flight, and then find the range using the horizontal component of the velocity.

Worked example

A cricket ball is struck so that the ball leaves the bat at 60° to the ground travelling at 25 m s^{-1}. Calculate:

(a) the maximum height reached by the ball (assume it starts at ground level)
(b) the total time the ball is in the air
(c) the horizontal distance from the bat when the ball first hits the ground

Answer

In the vertical plane:
$u = 25$ m s^{-1} $\cos 30 = 22$ m s^{-1}
$v = 0$ m s^{-1} (at the highest point, $s = h$)
$a = g = -9.8$ m s^{-2}

In the horizontal plane:
$u = 25$ m s^{-1} $\cos 60 = 12.5$ m s^{-1}

(a) $v^2 = u^2 + 2as$ $(0$ m s$^{-1})^2 = (22$ m s$^{-1})^2 - 2 \times 9.8$ m s$^{-2} \times h \rightarrow h = 8.0$ m
(b) $v = u + at$ 0 m s$^{-1} = 22$ m s$^{-1} - 9.8$ m s$^{-2} \times t \rightarrow t = 2.2$ s
    This is the time to the top of the trajectory, so the total time in the air will be 4.4 s.
(c) The ball travels at a constant horizontal velocity of 12.5 m s^{-1}
    horizontal displacement, $x = 12.5$ m s$^{-1} \times 4.4$ s $= 55$ m

Forces

Forces push, pull, squeeze or stretch. They fall into two categories: distant and contact forces.

- Forces acting over a distance: gravitational (e.g. between the Sun and the planets) and electromagnetic (between static and moving charges, and between magnetic poles). Nuclear forces are not considered until the A2 course.
- Contact forces: reaction forces between your shoe and the floor, friction, viscosity and air resistance.

Tension forces, such as the forces tending to restore the initial length of a stretched rubber band, are caused by short-range attractive forces between displaced molecules. These are electromagnetic in nature and are considered more fully under materials, later in the guide.

Newton's first law of motion

An object in outer space, if not subjected to any forces, will stay at rest or continue to move at constant velocity. Its motion will also be unaffected if any forces acting on the object cancel out. This principle was first suggested by Isaac Newton, and is now accepted as Newton's first law of motion.

Definition

- Newton's first law of motion states that a body will remain at rest or move with uniform velocity unless acted upon by a resultant external force.

Forces are vector quantities. You have seen that the addition of forces requires the direction of the forces to be considered as well as their magnitudes. The total sum of a number of forces, acting at a point, is termed the *resultant* force. When the resultant force acting on a body is zero, the body is said to be in *equilibrium.*

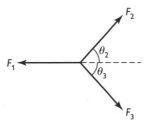

For P to be in equilibrium the vector sum of F_1, F_2 and F_3 must be zero. For this to be true, the sum of the horizontal components must be zero, and also those in the vertical plane. (In fact a body will be in equilibrium if the components of the forces in any two planes add up to zero, but it is most convenient to use components in planes at right angles.)

In the horizontal plane: $(F_2 \cos \theta_2 + F_3 \cos \theta_3) - F_1 = 0$

In the vertical plane: $F_2 \cos (90 - \theta_2) - F_3 \cos (90 - \theta_3) = 0$

Worked example

A sledge is being pulled across the snow at constant velocity by two husky dogs. One dog pulls with a force of 100 N at an angle of 20° to the direction of travel and the second animal pulls on its harness at an angle of 35° to that of the other dog.

Calculate:

(a) the force applied to the sledge by the second dog

(b) the resistive force of the snow on the sleigh

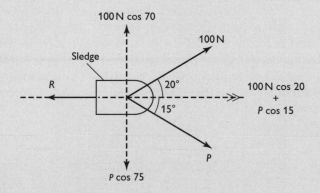

Answer

As the sledge is moving at constant velocity it is in equilibrium, i.e. the resultant of the forces of the dogs and the resistive force of the snow must be zero.

(a) In the plane perpendicular to the motion: 100 N cos 80 – P cos 75 = 0 →
$P = 67$ N

(b) In the direction of motion: (100 N cos 20 + 67 N cos 15) – R = 0 → R = 159 N

Free-body force diagrams

In the above example we considered only the horizontal forces acting on the sledge. In fact there are several other forces acting in the system; the weight of the sledge, the reaction of the ground on the sledge and dogs etc. When considering the equilibrium of a single body we need to isolate all the forces acting on it using a *free-body force diagram*.

Consider the trapeze artist shown in the diagram on p. 22. She 'feels' only three forces: her weight pulling down, the horizontal pull of her assistant and the trapeze rope through her hands.

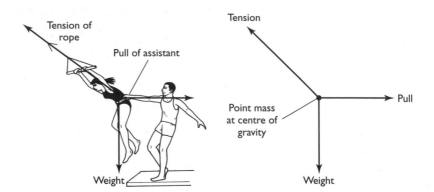

In order to use a free-body force diagram to find components and make calculations, the lines of action of the forces must be accurately drawn. It is clear that the pull of the rope is through the hands and arms and the assistant is holding the belt, but where is the weight acting? The diagram shows all the forces acting at a single point. For all extended bodies there is a point through which the weight always acts. This is known as the *centre of gravity* of the body. For free-body force diagrams it is recommended that the body is represented by a point at its centre of gravity, and all forces are drawn through this point.

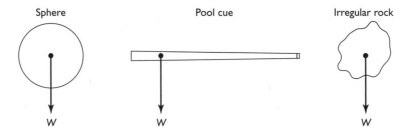

The position of the centre of gravity for three different objects is shown above. For regular objects like the sphere the centre of gravity is at the geometric centre, but its position in irregular bodies depends on the shape and the mass distribution. The centre of gravity will always be directly above the balance point (or vertically below the point of suspension) of the body. If you balance a pool cue on your finger and close your eyes, it would feel the same as a single point mass placed on your finger.

Newton's second law of motion

Newton's first law tells us that a body will remain at rest or continue to move with constant velocity if no resultant forces act on it. So what will happen if a resultant force is applied? There will be a change in its motion, its velocity will change, which means that the body will accelerate. For *fixed masses* the relationship between the resultant force and the acceleration is given by the expression:

$$\Sigma F = ma$$

This is an expression of Newton's second law of motion applied to bodies of fixed mass, and assumes the force to be in newtons, the mass in kilograms and the acceleration to be in metres per second squared. A more general definition of the law is used in Unit 4 of the A2 course, but only forces acting on fixed masses will be examined in Unit 1.

Tip Always remember to find the resultant force before applying the equation.

Worked example

An experiment to investigate the relationship between the acceleration of a fixed mass uses a glider on an air track and two optical timing gates as shown below.

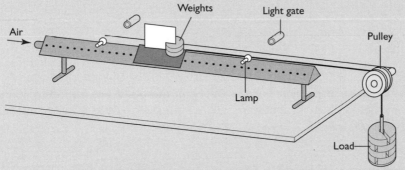

Investigating the acceleration of an object

(a) The light gates are connected to a computer that measures the times taken for the interrupter card to pass through each gate, and the time for the glider to travel from the first gate to the second one. Describe how you would perform the experiment, explaining how values of force and acceleration are obtained.

(b) A typical set of readings is given in the table. F represents the resultant force applied to the mass, t_1 and t_2 are the times for the card to cut through the gates, and t is the time taken by the glider to travel between the gates. Copy the table and complete the columns for the initial and final velocities, u and v, and for the acceleration, a. The length of the interrupter card is 20 cm.

F/N	t_1/s	t_2/s	t/s	u/m s^{-1}	v/m s^{-1}	a/m s^{-2}
0.10	1.25	0.48	2.00			
0.20	0.91	0.34	1.48			
0.30	0.74	0.28	1.18			
0.40	0.63	0.24	1.00			
0.50	0.57	0.23	0.84			

(c) Draw a graph of a/m s^{-2} against F/N. Use your graph to determine the accelerated mass, m.

(d) Calculate the acceleration of a car of mass 1500 kg if the driving force on the car is 1200 N and the resistive forces (friction, drag) total 450 N.

Answer

(a) The resultant force acting on the system (the glider plus all the masses) is the downward weight of the load. To increase this force a mass is taken off the glider and added to the load. Thus the total mass of the system stays the same.

The velocities of the glider at each gate, u and v, are found by dividing the length of the card by the times to cross the gates, and the acceleration is calculated using the equation $v = u + at$.

(b)

F/N	t_1/s	t_2/s	t/s	u/m s^{-1}	v/m s^{-1}	a/m s^{-2}
0.10	1.25	0.48	2.00	0.16	0.42	0.13
0.20	0.91	0.34	1.48	0.22	0.59	0.25
0.30	0.74	0.28	1.18	0.27	0.72	0.38
0.40	0.63	0.24	1.00	0.32	0.84	0.52
0.50	0.57	0.23	0.84	0.35	0.88	0.63

(c)

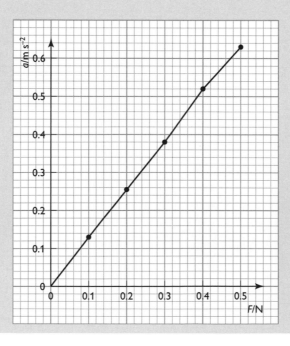

$$m = \frac{F}{a} = \frac{1}{\text{gradient}} = \frac{0.50 \text{ N}}{0.64 \text{ m s}^{-2}} = 0.78 \text{ kg}$$

(d) Resultant force acting on the car, $\Sigma F = 1200 \text{ N} - 450 \text{ N} = 750 \text{ N}$

$$\Sigma F = ma \rightarrow a = \frac{750 \text{ N}}{1500 \text{ kg}} = 0.5 \text{ m s}^{-2}$$

Unit of force — the newton

Up to this point in this guide we have just accepted the use of the newton (N) as the SI unit of force. Newton's second law equation for a fixed mass requires the mass to be in kilograms, the acceleration in metres per second squared and the resultant force to be in newtons. It follows from the equation that a resultant force of 1 newton acting on a mass of 1 kilogram will accelerate the mass at a rate of 1 metre per second squared.

Definition

- 1 newton (N) is that resultant force which, when acting on a mass of 1 kilogram, produces an acceleration of 1 metre per second squared.

Gravitational field strength

The section relating to free-falling bodies using the ideas of Galileo and subsequent experimental evidence showed that objects accelerate at 9.8 m s^{-2} when released close to the Earth's surface (ignoring air resistance). Applying Newton's second law equation to a mass, m, close to the surface:

$$\text{Acceleration due to gravity, } g = \frac{F}{m} = 9.8 \text{ m s}^{-2}$$

The concept of a field in physics relates to a region where a force is experienced. A gravitational field is therefore a region where a gravitational force is felt, i.e. a mass placed in the field will experience a force. The strength of a gravitational field will determine the size of the force acting upon a given mass. Hence the above expression for the acceleration due to gravity also acts as a measure of gravitational field strength.

$$\text{Gravitational field strength, } g = \frac{F}{m} = 9.8 \text{ N kg}^{-1}$$

The gravitational field strength on the surface of the Earth varies a little. It is slightly greater in the polar regions than on the Equator and can be affected by high-density metallic ores etc. The value of g also decreases with height above the surface such that a rocket reaching 50 000 km will experience little gravitational force from the Earth. The gravitational field strength on the Moon is about 1.6 N kg^{-1}, one-sixth of that on the Earth.

Weight is the force acting upon a mass in a gravitational field. Rearranging the expression for the field strength:

$$W = m\,g$$

So a mass of 1 kg has a weight of 9.8 N on Earth and a weight of 1.6 N on the Moon.

Newton's third law of motion

Definition
- Newton's third law of motion states that if a body A exerts a force on a body B, then body B will exert an equal and opposite force on body A.

It must be stressed that this law cannot be applied to single bodies, and you should always state which bodies the forces act upon and the direction of the forces.

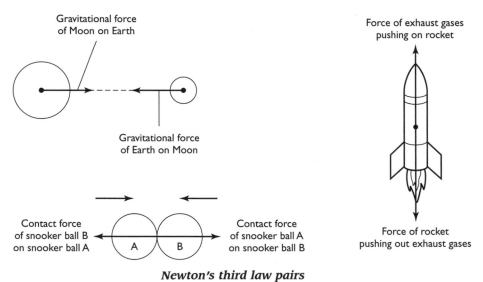

Newton's third law pairs

The examples shown above illustrate Newton's third law. Each shows a pair of forces acting on two different bodies. These pairs must always:
- act on two separate bodies
- be of the same type
- act along the same line
- be equal in magnitude
- act in opposite directions

Sometimes there can be more than one pair of forces acting on two bodies. Consider a woman standing on the Earth:

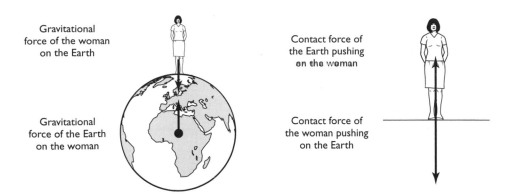

Newton's third law pairs of forces acting between the Earth and a woman

The diagram shows the two Newton's third law pairs acting on the woman and the Earth:

- gravitational force of the Earth on the woman (down); gravitational force of the woman on the Earth (up)
- contact (elastic) force of the woman pushing down on the floor; contact force of the floor pushing up on the woman

If the woman alone is considered, it can be seen that she is in equilibrium due to the action of the two different types of force.

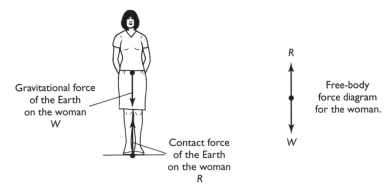

Free-body force diagram for the woman

The resultant of the upward contact force and the downward gravitational force (her weight) is zero, and so she will remain at rest according to Newton's first law.

> **Tip** It is a common error to regard these two forces as a third law pair, so you must remember that two separate bodies are needed for Newton's third law.

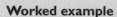

Worked example

A fisherman checks the weight of his catch using a newton-meter. Two of the forces acting are shown in the diagram below.

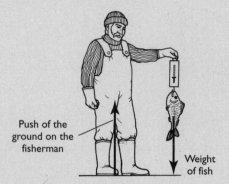

Push of the ground on the fisherman

Weight of fish

(a) For each force there is a 'Newton's third law pair force'. In each case state:

    (i) the body that the Newton's third law pair force acts upon

    (ii) the type of force

    (iii) the direction of the Newton's third law pair force

(b) Draw a free-body diagram for the fish, labelling the forces acting upon it.

Answer

(a)

Force	(i)	(ii)	(iii)
Weight of the fish	The Earth	Gravitational	Upwards
Push of the ground on the fisherman	Push of the fisherman on the ground	Contact (elastic) force	Downwards

(b)

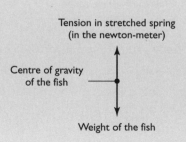

Tension in stretched spring (in the newton-meter)

Centre of gravity of the fish

Weight of the fish

Work, energy and power

Work and energy are interrelated. When work is done on a body, the body will gain energy, and a body can transfer energy to do work. Power is the rate of doing work.

Work

- Work is done when the point of application of a force is moved.
- Work done equals the product of the force *times* the distance moved *in the direction of the applied force.*
- $\Delta W = F\Delta x$ is the equation defining the work done when the point of application of a force, F, is moved a distance Δx.
- 1 joule (J) of work is done when the point of application of a force of 1 newton (N) is moved through a distance of 1 metre (m).

Worked example

Calculate the work done in lifting a load of 8.0 kg onto a shelf of height 1.6 m.

Answer

Force needed to raise the load, $F = mg = 8.0 \text{ kg} \times 9.8 \text{ m s}^{-2} = 78 \text{ N}$

Work done $= F\Delta x = 78 \text{ N} \times 1.6 \text{ m} = 125 \text{ J}$

Tip Most questions involving objects to be lifted give the mass of the object. It is a common error to use this as the weight. Remember always to multiply the mass by 9.8 m s^{-2} in order to find the weight of a body.

The definition of work requires that the distance moved is in the direction of the point of application of the force. In some cases a body will move in a different direction to the applied force. Consider a sleigh being pulled across the snow by a piece of rope.

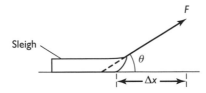

The direction of movement is at an angle θ to the applied force, so the work done will be given by:

$$\Delta W = F \cos \theta \times \Delta x = F \Delta x \cos \theta$$

You might think from the above expression that work is a vector quantity. However, if the sleigh is being pulled along a curved path, work is continually being done on it in ever-changing directions, and so work is a scalar quantity.

Energy

- Energy is the capacity to do work.
- Energy is a scalar quantity, measured in joules.
- Energy cannot be created or destroyed, but it can be converted to another form or transferred as work.

The relationship between work and energy and the law of conservation of energy are fundamental principles in physics at all levels. You will probably have considered the energy changes in simple systems such as a light bulb connected to a battery (chemical energy → electrical energy → heat and light energy) or even the more complex system of a gas-fired power station (chemical energy in the gas → heat energy → kinetic energy of the turbine → electrical energy in the generator). In this unit you will be required to look more deeply into the mechanical forms of energy: *potential energy* and *kinetic energy*.

Gravitational potential energy (GPE)

If a mass, m is raised from the floor through a height Δh the work done on the mass is:

$$\Delta W = mg\Delta h$$

The mass is now in a position to do work by falling back to the floor. The mass has potential energy due to its position.

Definition

- Gravitational potential energy is the ability of a body to do work by virtue of its position in a gravitational field
- $\Delta \mathbf{GPE} = \mathbf{\textit{mg}\Delta \textit{h}}$

Elastic potential energy (EPE)

A stretched rubber band can do work when it is released. The band is therefore storing energy while it is in its deformed state. This energy is called **elastic potential energy** (EPE) or **elastic strain energy**. Any elastically deformed object will store EPE. Clockwork springs, loaded crossbows and catapults (slingshots) are all examples of systems storing elastic potential energy.

Elastic strain energy will be covered in more detail later in this guide, in the section on solid materials (p. 49).

Kinetic energy (KE)

If a cyclist freewheeling along a level road applies the brakes, the frictional resistive force between the brake blocks and the wheel will bring the cycle to rest over a distance depending on its speed, the nature of the brakes etc. The moving cycle and rider will therefore be doing work against the resistive forces. The moving cycle must have some energy to do this work; this energy is called **kinetic energy**.

Definition

- Kinetic energy is the ability of a body to do work by virtue of its motion.
- **KE = ½mv^2**

In the case of the cyclist, the kinetic energy is transferred to the brake blocks as thermal energy that is dissipated into the environment.

Worked example

Roller-coaster ride

The car on a roller-coaster ride falls through a height of 20 m from rest at the top of the first peak to the bottom of the first dip. At the top of the second peak, 12 m above the dip, it is travelling at 12 m s⁻¹. The mass of the car and passengers is 2000 kg.

(a) Show that the speed of the car in the dip is about 20 m s⁻¹.

(b) Estimate the work done by the car against the resistive forces between the two peaks.

Answer

(a) loss in GPE = gain in KE

$$mg\Delta h = ½mv^2 \quad v = \sqrt{(2gh)} \quad = \sqrt{(2 \times 9.8 \text{ m s}^{-2} \times 20 \text{ m})} = 19.8 \text{ m s}^{-1}$$

> **Tip** In a 'show that' question the answer should be given to at least one more significant figure than the approximate value shown in the question.

(b) loss in GPE = gain in KE + work done against resistive forces

$$2000 \text{ kg} \times 10 \text{ m s}^{-2} \times (20 \text{ m} - 12 \text{ m}) = ½ \times 2000 \text{ kg} \times (12 \text{ m s}^{-1})^2 + \Delta W$$

$$\Delta W = 160 \text{ kJ} - 144 \text{ kJ} = 16 \text{ kJ} \approx 20 \text{ kJ}$$

Note that for an estimate it is acceptable to use $g = 10$ m s⁻² and give the answer to one significant figure.

Power

- Power is the rate of doing work, or transferring energy.
- $P = \dfrac{\Delta W}{\Delta t} = \dfrac{\Delta E}{\Delta t}$

Power is measured in watts, W (J s^{-1}).

Worked example

A man of mass 75 kg runs up a flight of stairs with 20 steps each of height 20 cm in a time of 4.0 s. Calculate the average power expended by the man.

Answer

Work done = $mg\Delta h$ = 75 kg × 9.8 m s^{-2} × (20 × 0.20 m) = 1470 J

$$\text{Power} = \frac{1470\ \text{J}}{4.0\ \text{s}} = 370\ \text{W}$$

The power of an object moving at constant velocity against a uniform resistive force can be found from the product of the force and the velocity:

$$P = \frac{\Delta W}{\Delta t} = F\frac{\Delta x}{\Delta t} = F \times \frac{\Delta x}{\Delta t} = F \times v$$

Efficiency

Earlier in this section the energy conversions for a light bulb and a power station were given. In both these cases much of the initial energy input is not converted to the desired output. For a filament lamp most of the electrical energy is converted to heat leaving a relatively small percentage as light. Similarly, throughout the conversions in the power station, and particularly in the turbines, a large amount of heat is 'lost' to the surroundings. The efficiency of a system relates to how much useful output is gained compared with the energy put into the system.

$$\text{Efficiency} = \frac{\text{useful energy (power) output}}{\text{energy (power) input}} \times 100\%$$

Worked example

An invalid buggy is powered by a 300 W electric motor. It moves with a speed of 2.5 m s^{-1} against an average resistive force of 100 N.
Calculate the output power, and hence the efficiency of the buggy.

Answer

Output power = $F \times v$ = 100 N $\times$ 2.5 m s^{-1} = 250 W

$$\text{Efficiency} = \frac{250 \text{ W}}{300 \text{ W}} \times 100\% = 83\%.$$

Materials

Properties of fluids

A fluid is a material that flows. Unlike a solid, in which the atoms occupy fixed positions, the particles of a fluid can move relative to each other. For this unit fluids will be considered as liquids and gases only, but you should be aware that plasma in stars and some solids such as glass also exhibit fluid characteristics.

Density of fluids

- Density $= \dfrac{\text{mass}}{\text{volume}}$ $\quad \rho = \dfrac{m}{v}$ units: kg m^{-3}

Fluid	Density/kg m^{-3}	Fluid	Density/kg m^{-3}
Mercury	13 600	Carbon dioxide	1.78
Sulfuric acid in car battery	1 250	Air	1.24
Water	1 000	Helium	0.161
Ethanol	790	Hydrogen	0.081

The table shows the wide range of in density of fluids. Gases have much lower densities than liquids because the molecular separation is much larger. Liquids are virtually incompressible, but the density of gases will increase with increasing pressure. The values in the table are for pressures of 1.01 $\times$ 10^5 Pa at 293 K. The relationships between pressure, temperature and volume of gases will be studied in more detail in Unit 5 of the A2 course.

Worked example

In an experiment to measure the density of air, the mass of a glass flask is taken before and after removing most of the air using a vacuum pump. The volume of the air removed at normal pressure is found by putting the end of the tubing into a beaker of water and releasing the clamp.

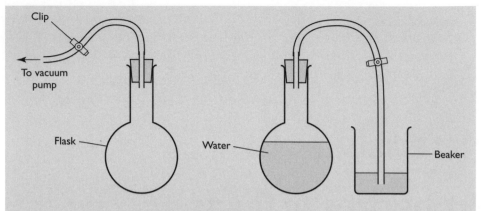

The measurements taken are given below:

    Mass of flask plus air = 212.32 g

    Mass of flask after pumping = 211.35 g

    Volume of water drawn into flask = 785 ml

Use these measurements to determine the density of air.

Answer

Mass of air removed = 212.32 g – 211.35 g = 0.97 g = 9.7×10^{-4} kg

Volume of air removed = 785 ml = 7.85×10^{-4} m³

$$\text{Density} = \frac{9.7 \times 10^{-4} \text{ kg}}{7.85 \times 10^{-4} \text{ m}^3} = 1.24 \text{ kg m}^{-3}$$

Pressure in fluids

At any point in a column of fluid there is a pressure that acts equally in all directions, the value of which depends on the height of the fluid above that point.

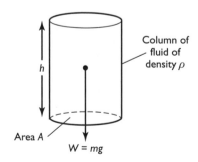

$$\text{Pressure} = \frac{\text{force}}{\text{area}} = \frac{\text{weight of column}}{\text{area}} = \frac{mg}{A} = \frac{V\rho g}{A} = \frac{(Ah)\rho g}{A} = h\rho g$$

> **Worked example**
>
> Estimate the height of the Earth's atmosphere given that the density and air pressure at the surface are 1.24 kg m^{-3} and 1.01 × 10^5 Pa respectively.
>
> *Answer*
>
> Assuming the average density to be about 0.6 kg m^{-3} and taking $g = 10$ m s^{-2}.
>
> 1×10^5 Pa $= h \times 0.6$ kg m$^{-3} \times 10$ m s$^{-2} \rightarrow h = 17$ km ≈ 20 km

It is important to be aware of the effect of the density of the fluid in pressure calculations. At one time, atmospheric pressure was measured in terms of the height of the column of mercury that balanced the air pressure using a mercury barometer. The standard value was 760 mm of mercury. Similarly the pressure at a depth of 10 m under water will give the same value. Use the data in the table of densities (on p. 33) to check that both produce a pressure of 1 × 10^5 Pa.

Upthrust

If you are in a swimming pool, you will experience a buoyancy force that enables you to float or swim. This force is called an *upthrust*, and it is a consequence of the water pressure being greater below an immersed object than above it.

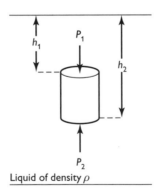

Liquid of density ρ

- pressure on the upper surface, $P_1 = h_1 \rho g$
- pressure on the lower surface, $P_2 = h_2 \rho g$
- $\Delta P = (h_2 - h_1)\rho g$
- upthrust, $\Delta F = \Delta P \times A = (h_2 - h_1)\rho g \times A = (h_2 - h_1)A\rho g = V\rho g$
- $\Delta F = mg = $ *weight of the fluid displaced*

This result is often stated as *Archimedes' Principle,* which tells us that the upthrust on a body immersed in a fluid equals the weight of the displaced fluid. It follows that, for a given volume of displaced fluid, the upthrust is proportional to the density of the fluid. You will experience greater flotation forces in sea water than in fresh water because brine has a larger density. Also, the upthrust of the air on your body is so small that it is not noticeable as the density of the air is very small.

Tip In many upthrust calculations you will be given the mass of displaced fluid. Make sure that this is converted to weight by multiplying by 9.8 m s^{-2}.

Worked example

An uninflated toy balloon has a mass of 4.9 g. It is filled with helium to form a sphere of diameter 24 cm, and is held by a length of string tied to a post so that it floats directly above the point of attachment. Use the data in the fluid density table (p. 33) to show that the tension in the string is about 0.03 N.

Answer

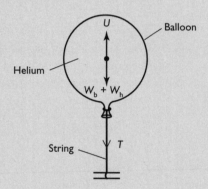

For equilibrium:

$U = W + T$

U = weight of displaced air = $4/3 \, \pi \, (0.12 \text{ m})^3 \times (0.161 \text{ kg m}^{-3}) \times 9.8 \text{ m s}^{-2} = 0.088 \text{ N}$

W = weight of rubber + weight of helium

$\quad = W_b + W_h$

$\quad = 4.9 \times 10^{-3} \text{ kg} \times 9.8 \text{ m s}^{-2} + 4/3 \, \pi \, (0.12 \text{ m})^3 \times (1.24 \text{ kg m}^{-3}) \times 9.8 \text{ m s}^{-2}$

$\quad = 0.059 \text{ N}$

$T = 0.088 \text{ N} - 0.059 \text{ N} = 0.029 \text{ N} \approx 0.03 \text{ N}$

Moving fluids

In fluid flow, particles within the fluid are affected by *cohesive* forces of neighbouring molecules and *adhesive* forces to the surfaces of obstructing solids or the inner walls of a pipe. The path taken by an individual particle within a moving fluid is called a **streamline**.

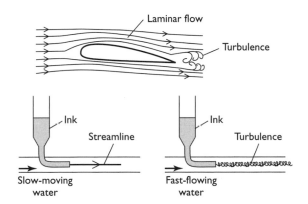

When the streamlines of adjacent particles do not cross over each other the flow is said to be **laminar** (often called **streamlined flow**). When the paths do cross the flow becomes **turbulent**.

The above diagram shows that for an aircraft wing the airflow needs to be laminar around the wing, although some turbulence is likely beyond the wing. For fluids in a tube the flow will be streamlined at low rates, becoming turbulent at some critical velocity when the fluid swirls around forming **vortices** or **eddy currents**.

Viscosity

When a fluid moves through a pipe, or flows relative to a solid body, it will experience resistive forces that impede its motion. The rate of flow depends upon the **viscosity**, the degree of 'stickiness', of the fluid. For example treacle and most oils will flow much less readily than water, and gases generally will have a much lower viscosity than liquids.

Relative values of the viscosities of liquids can be found using a simple **viscometer** as shown in the diagram on p. 38.

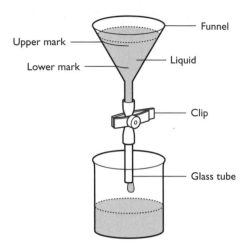

The time taken for the liquid level in the reservoir to fall between the two marks is taken for a range of different liquids.

The rate of flow of a fluid though a tube depends upon:
- the viscosity of the fluid
- the diameter of the tube
- the length of the tube
- the pressure across its ends
- whether the flow is streamlined or turbulent

The most significant factor is the diameter of the tube. If the diameter is doubled the flow rate increases sixteen times if all other factors remain the same. The diameters of oil and gas pipes need to be as large as possible to enable an economic flow rate. Restriction in blood flow in narrowed arteries is a common cause of high blood pressure.

It is also important to be aware that the viscosity of most fluids will decrease at higher temperatures.

Stokes's law

Stokes's law relates to the special case of a spherical body falling through a fluid in which the flow of the fluid relative to the body is streamlined.

Definition
- Viscous drag, $F = 6\pi\eta rv$
- η is the coefficient of viscosity (N s m^{-2})
- r is the radius of the sphere
- v is the relative velocity of the fluid to the sphere

In addition to the motion of solid spheres in liquids, Stokes's law is used in the study of raindrops, mist particles and aerosol droplets in the atmosphere.

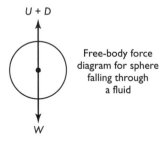

$U + D$

Free-body force diagram for sphere falling through a fluid

W

The free-body diagram for a sphere falling through a fluid shows that the resultant force acting on the sphere will be:

$F = W - (U + D)$

Initially, when the sphere is stationary, the viscous drag is zero and the resultant downward force equals the weight minus the upthrust. The sphere accelerates downwards and, as the velocity increases, so does the viscous drag. The resultant force is therefore reduced until it becomes zero, the sphere is in equilibrium, and it stops accelerating. The sphere will then continue to fall with a constant velocity known as its **terminal velocity**.

Measuring the terminal velocity of a sphere falling through a fluid provides a method of obtaining a value for the viscosity of the fluid.

When the terminal velocity is reached:

$W - (U + D) = 0 \quad W = U + D$

$\frac{4}{3} \pi r^3 \rho_s g = \frac{4}{3} \pi r^3 \rho_f g + 6\pi\eta rv$

(weight = mg = volume of sphere × density × g; ρ_s and ρ_f are the densities of the solid and the fluid)

$$\eta = \frac{2(\rho_s - \rho_f)gr^2}{9v}$$

Worked example

In an experiment to determine the viscosity of motor oil at room temperature, a steel ball-bearing is released just below the surface of the oil and timed as it crosses a series of equally spaced marks on the glass cylinder. The experiment is repeated three times, and the average time, Δt, taken for the ball to cross each 5 cm division is calculated.

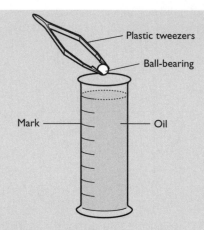

The results of the experiment are given below:

  Separation of marks = 5.0 cm

  Diameter of ball bearings = 3.0 mm

  Density of steel = 7800 kg m^{-3}

  Density of oil = 820 kg m^{-3}

(a) Copy and complete the table to include the incremental times and the average velocity between successive marks.

Mark	Average t/s	Δt/s	v/m s^{-1}
1	1.43		
2	2.52		
3	3.40		
4	4.31		
5	5.22		
6	6.13		

(b) Sketch a graph to show how the velocity of the ball changes as it falls through the oil.

(c) Determine the value of the critical velocity, and hence calculate the viscosity of the oil.

Answer

(a)

Mark	Average t/s	Δt/s	$v \times 10^{-2}$/m s^{-1}
1	1.43	1.43	3.50
2	2.52	1.09	4.59
3	3.40	0.98	5.10
4	4.31	0.91	5.49
5	5.22	0.91	5.49
6	6.13	0.91	5.49

(b)

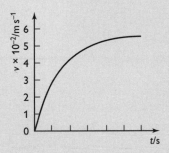

(c) Terminal velocity $v = 5.5 \times 10^{-2}$ m s^{-1}

$$\eta = \frac{2(\rho_s - \rho_f)gr^2}{9v}$$

$$= \frac{2 \times (7\,800 - 820)\ \text{kg m}^{-3} \times 9.8\ \text{m s}^{-2} \times (1.5 \times 10^{-3}\ \text{m})^2}{9 \times 5.5 \times 10^{-2}\ \text{m s}^{-1}} = 0.61\ \text{N s m}^{-2}$$

Viscosity of gases

The coefficient of viscosity of gases is much lower than that of liquids, and so the effect is less apparent unless the flow rates are high. For domestic gas boilers close to the supply, a short length of 15 mm diameter copper pipe is sufficient for an adequate flow, but for appliances requiring several metres of tubing with some bends, regulations stipulate the use of 22 mm tube.

Because of the low viscosity of air (about 2×10^{-5} N s m^{-2}) the terminal velocity of falling objects is generally very large, but it depends a great deal on the size of the body. For bodies falling through air we can usually ignore the upthrust, because the air density is so small, so an object falling from rest will initially experience no upward forces and will begin to accelerate downwards at 9.8 m s^{-2}. As the speed increases the viscous drag also increases. A velocity–time graph similar to the one in the above example will represent this motion. The initial gradient will equal g and the terminal velocity is the value when the curve has levelled off.

Worked example

Estimate the terminal velocity of a raindrop of diameter 1 mm and a droplet of mist of diameter 0.1 mm (viscosity of air = 2×10^{-5} N s m^{-2}, density of water = 1000 kg m^{-3}).

Answer

Ignoring the upthrust, at equilibrium:

Weight of drop = viscous drag

$$\frac{4}{3}\pi r^3 g = 6\pi r v \quad \rightarrow v = \frac{2r^2 \rho g}{9\eta}$$

For 1 mm diameter $v = \dfrac{2 \times (0.5 \times 10^{-3} \text{ m})^2 \times 1000 \text{ kg m}^{-3} \times 10 \text{ m s}^{-2}}{9 \times 2 \times 10^{-5} \text{ N s m}^{-2}} = 30 \text{ m s}^{-1}$

For 0.1 mm diameter $v = \dfrac{2 \times (0.05 \times 10^{-3} \text{ m})^2 \times 1000 \text{ kg m}^{-3} \times 10 \text{ m s}^{-2}}{9 \times 2 \times 10^{-5} \text{ N s m}^{-2}} = 0.3 \text{ m s}^{-1}$

In nature, the larger drops actually fall at about 5 m s^{-1}, but the speed of the smaller drops is about right. It follows that Stokes's law applies to the slower small droplets but not to the larger ones.

Tensile behaviour of solid materials

Tensile means linear extensions or compressions due to applied forces. The behaviour of many materials subjected to tensile forces is similar to that of a spring.

Hooke's law

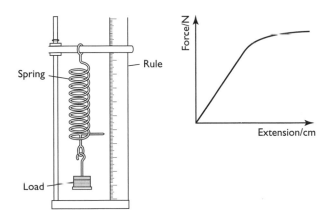

A graph of force against extension shows that the extension is directly proportional to the load until the spring is overloaded and begins to lose its elasticity. Within the region of proportionality the spring obeys **Hooke's law**.

- Hooke's law states that, up to a certain limit, the extension is directly proportional to the load
- $F = k\Delta x$
- k is the *spring constant* $= \dfrac{F}{\Delta x}$ N m^{-1} = gradient of the linear region

Any material that extends (or compresses) proportionately to the applied load obeys Hooke's law. You will see later that most metals obey the law for relatively low loads, but many polymers do not, whatever the load.

Force–extension graphs

The properties of solid materials are readily illustrated using force–extension graphs. An experiment in which a long piece of copper wire is loaded until it breaks, taking corresponding values of load and extension produces a force–extension graph that is typical of most metals.

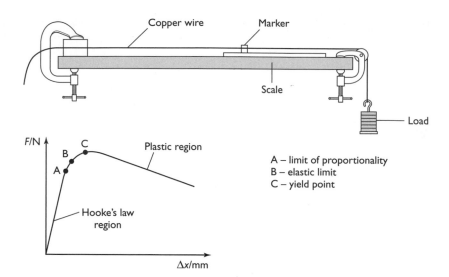

A – limit of proportionality
B – elastic limit
C – yield point

The following regions or points need to be known and described in the examination:

- **Hooke's law region** — where F is proportional to Δx.
- **Limit of proportionality** — point where Hooke's law ceases to be obeyed.
- **Elastic limit** — point beyond which the wire will not regain its original length when the load is removed.
- **Plastic flow** — the region where the metal deforms plastically. If the load is removed there will be little or no change in the length of the wire.
- **Yield point** — the onset of plastic flow.

When a metal behaves *elastically*, the bonds holding the atoms together are stretched. When the force is removed the atoms return to their original positions and the metal regains its shape.

At the yield point, layers of atoms slide over one another and are unable to return to the initial positions. This is termed **plastic flow**.

Force–compression graphs are more difficult to obtain, unless specialised equipment such as screw or hydraulic presses is available. The elastic region is much the same as the extension graphs, but after the Hooke's law region the metal rods tend to buckle and break.

Polymers such as plastics and rubber have more complex molecular structures than metals and so behave differently when deformed. If the spring in the earlier experiment was replaced by a rubber band, the force–extension graph shown below would be plotted.

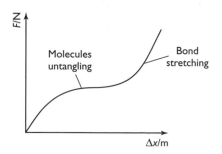

Polymers have long chain molecules that are initially in a 'tangled' state, a little like a bowl of cooked spaghetti. When a tensile force is applied the 'cross-links' between the polymers are broken and the chains straighten out, leading to a large extension for a relatively small additional load. When the polymers become untangled, the load has to stretch the bonds and so the rubber becomes much more difficult to stretch.

Stress, strain and the Young modulus

Force–extension graphs give us information relating to a particular specimen of a solid material, but these curves vary for different sized samples of the same material. In order to make the graphs consistent for a given material, the dimensions must be taken into account. In order to do this, the following quantities are defined for materials subjected to tensile forces:

- **Stress** = $\dfrac{\text{tensile force}}{\text{cross-sectional area}}$ $\qquad \sigma = \dfrac{F}{A}$ $\qquad$ unit: pascal Pa (N m^{-2})

- **Strain** = $\dfrac{\text{extension}}{\text{original length}}$ $\qquad \varepsilon = \dfrac{\Delta x}{l}$ $\qquad$ no unit

- **Young modulus** = $\dfrac{\text{stress}}{\text{strain}}$ $\qquad E = \dfrac{\sigma}{\varepsilon}$ $\qquad$ unit: Pa

Worked example

A steel wire of length 2.00 m and diameter 0.40 mm is extended by 4.0 mm when a tensile force of 50 N is applied. Calculate:

(a) the stress
(b) the strain
(c) the Young modulus of steel

Answer

(a)

$$\text{Stress, } \sigma = \frac{50 \text{ N}}{\pi \,(0.20 \times 10^{-3} \text{ m})^2} = 4.0 \times 10^8 \text{ Pa}$$

(b)

$$\text{Strain, } \varepsilon = \frac{0.40 \times 10^{-3} \text{ m}}{2.00 \text{ m}} = 2.0 \times 10^{-3} \,(= 0.20\%)$$

(c)

$$\text{Young modulus, } E = \frac{4.0 \times 10^8 \text{ Pa}}{2.0 \times 10^{-3}} = 2.0 \times 10^{11} \text{ Pa}$$

Stress–strain graphs

The shape of the stress–strain graphs for metals and polymers is much the same as the force–extension graphs. For relatively small extensions of a long wire, the change in cross-sectional area is very small, so the stress is approximately proportional to the load, and the strain is always proportional to the extension.

Stress–strain graphs have the advantage of being applicable to the material under test, and not just a particular sample. In the Hooke's law region, where stress is proportional to strain, the gradient of the line equals the Young modulus of the material.

The **ultimate tensile stress** (UTS) or breaking stress is the maximum stress a material can bear before it breaks. It is a surprising fact that the silk of a spider's web has a greater UTS than steel.

Worked example

The results of an experiment in which a length of copper wire is loaded until it breaks are given below:

Length of wire = 3.00 m   Mean diameter of wire = 0.52 mm

(a) Complete the table to include the values of stress and strain.

Force/N	Extension/mm	Stress/Pa	Strain
0	0		
5.0	0.5		
10.0	1.0		
15.0	1.5		
20.0	2.0		
25.0	2.5		
30.0	3.0		
35.0	4.0		
40.0	6.0		
35.0	300		

(b) Plot a stress–strain graph for copper.
(c) Use your graph to determine the Young modulus and the UTS of copper.

content guidance

Answer

(a)

Force/N	Extension/mm	Stress × 10⁷/Pa	Strain × 10⁻⁴
0	0	0	0
5.0	0.5	2.4	1.7
10.0	1.0	4.7	3.3
15.0	1.5	7.1	5.0
20.0	2.0	9.4	6.7
25.0	2.5	11.8	8.3
30.0	3.0	14.2	10
35.0	4.0	16.5	13
40.0	6.0	18.9	20
35.0	300	16.5	100

(b)

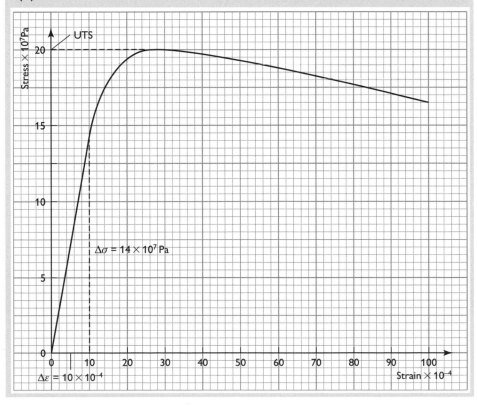

(c)

$$\text{Young modulus} = \text{gradient of linear region} = \frac{14 \times 10^7 \text{ Pa}}{10 \times 10^{-4}} = 1.4 \times 10^{11} \text{ Pa}$$

$$\text{UTS} = 2.0 \times 10^8 \text{ Pa}$$

Properties of solid materials

In this section we have explored the tensile behaviour of solid materials. You also need to be able to explain the meanings of a range of properties and describe how these are used in a variety of applications. The following table lists the properties, together with their descriptions or definitions, the name of the opposite property and some examples of materials that exhibit the property.

Property	Definition	Example	Opposite	Definition	Example
Strong	High breaking stress	Steel	Weak	Low breaking stress	Expanded polystyrene
Stiff	Gradient of a force–extension graph High Young modulus	Steel	Flexible	Low Young modulus	Natural rubber
Tough	High energy density up to fracture: metal that has a large plastic region	Mild steel, copper, rubber tyres	Brittle	Little or no plastic deformation before fracture	Glass, ceramics
Elastic	Regains original dimensions when the deforming force is removed	Steel in Hooke's law region, rubber	Plastic	Extends extensively and irreversibly for a small increase in stress beyond the yield point	Copper, plasticine
Hard	Difficult to indent the surface	Diamond	Soft	Surface easily indented/scratched	Foam rubber, balsa wood
Ductile	Can be readily drawn into wires	Copper	Hard, brittle		
Malleable	Can be hammered into thin sheets	Gold	Hard, brittle		

Elastic strain energy

In the energy section, elastic strain energy was defined in terms of the *potential* or stored energy in an elastically deformed body.

For a tensile extension:

Elastic strain energy = work done in extending the wire

= average force × extension

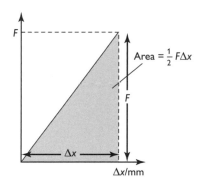

If Hooke's law is obeyed:

- $E_{el} = \frac{1}{2}F\Delta x$ = area under the line on the force–extension graph
- $E_{el} = \frac{1}{2}(k\Delta x) \times \Delta x = \frac{1}{2}k(\Delta x)^2$

It can be shown that the area under the line of any force–extension graph, even if Hooke's law is not followed, represents the elastic strain energy in the material.

For stress–strain graphs the area under the line represents the elastic strain energy per unit volume, commonly called the *energy density* of the deformed material.

Hysteresis in rubber

If a rubber band is first loaded and then unloaded, by adding and then removing a number of equal masses, and extensions are measured at each increment, the resulting force–extension graph will be as shown on p.50.

The graph shows that the unloading curve lies beneath the loading curve. This is known as **hysteresis** and can be explained in terms of elastic strain energy:

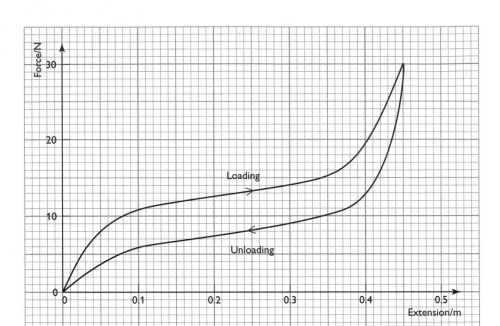

- During loading, work is done on the rubber. This is stored as elastic strain energy and is represented by the area under the loading curve.
- During unloading, the energy stored in the rubber does work by raising the load. This is represented by the area under the unloading curve.
- More energy has been used in stretching the band than has been retrieved during unloading. By the law of conservation of energy this 'lost' energy must be transferred to another form. This will be in the form of internal energy within the rubber, which will increase the temperature of the band and then be dispersed as thermal energy to the surroundings.
- Increase in internal energy within the rubber during one loading–unloading cycle = area enclosed by the loop.

Worked example

Use the hysteresis graph for rubber to estimate the increase in internal energy during the cycle shown.

Answer

Number of squares enclosed in the loop is about 8
The area of each square represents 5 N × 0.05 m = 0.25 J
Internal energy = 8 × 0.25J = 2 J

Other methods of estimating the area (e.g. considering each curve as two triangles and a trapezium and subtracting the lower area from the upper one) can be used if preferred.

Questions
&
Answers

The following two tests are made up of questions similar in style and content to the Unit 1 examination. Each paper carries a total of 80 marks and should be completed in 1 hour and 20 minutes. The first 10 questions are multiple-choice objective tests with four alternative responses. The remaining questions are varied in length and style, and carry marks ranging from 3 to about 12 marks. You may like to attempt a complete paper in the allotted time and then check your answers, or maybe do the multiple-choice section and selected questions to fit your revision plan. It is worth noting that there are 80 marks available for the 80 minute test, so this should help in determining how long you should spend on a particular question. You should therefore be looking at about 10 minutes for the multiple-choice section and approximately a minute a mark on the others.

You should also be aware that during the examination you must write your answers directly onto the paper. This will not be possible for the tests in this book, but the style and content are the same as the examination scripts in every other respect. It may be that diagrams and graphs that would normally be added to on the paper have to be copied and redrawn. If you are doing a timed practice test, you should add an extra few minutes to allow for this.

The answers should not be treated as model answers because they represent the bare minimum necessary to gain the marks. In some instances, the difference between an A-grade response and a C-grade response is suggested. This is not possible for the multiple-choice section, and many of the shorter questions do not require extended writing.

Ticks (✓) are included in the answers to indicate where the examiner has awarded a mark. Half marks are not given.

Examiner's comments

Where appropriate, the answers are followed by examiner's comments, denoted by the icon *e*. They are interspersed in the answers and indicate where credit is due and where lower-grade candidates typically make common errors. They may also provide useful tips.

Test Paper 1

Time allowed: 1 hour 20 minutes. Answer ALL the questions.

Questions 1–10

For questions 1–10 select one answer from A to D.

1 **A car accelerates uniformly from 10 m s^{-1} to 16 m s^{-1} in a time of 1 minute 40 seconds. The value of the acceleration during this time is:**
  A 0.01 m s^{-2} **B** 0.06 m s^{-2} **C** 0.10 m s^{-2} **D** 0.16 m s^{-2} (1 mark)

2 **The total distance travelled by the car in question 1 is:**
  A 0.50 km   **B** 1.0 km   **C** 1.3 km   **D** 1.6 km   (1 mark)

3 **Which of the following is not a vector quantity?**
  A acceleration   **B** displacement   **C** potential energy   **D** velocity   (1 mark)

4 **A golf ball leaves the club with a velocity of 30 m s^{-1} at an angle of 30° to the ground. The horizontal component of the velocity is:**
  A 10 m s^{-1} **B** 14 m s^{-1} **C** 20 m s^{-1} **D** 26 m s^{-1} (1 mark)

5 **The units of *tensile strain* are:**
  A N m^{-1} **B** N m^{-3} **C** no unit   **D** Pa   (1 mark)

6 **A body will always continue to travel with constant velocity if:**
  A a uniform resultant force is acting upon it
  B it is in equilibrium
  C it is in a vacuum
  D two forces of equal magnitude act upon it   (1 mark)

Questions 7 and 8 relate to the following force–extension graph.

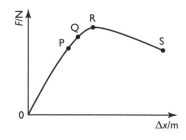

7 **Hooke's law is obeyed in the region:**
  A 0–P   **B** 0–Q   **C** 0–R   **D** R–S   (1 mark)

8 **The elastic limit is most likely to be at:**
  A P   **B** Q   **C** R   **D** S   (1 mark)

In questions 9 and 10, which of the following velocity–time graphs best represents the situation described? Each graph may be used once, more than once or not at all.

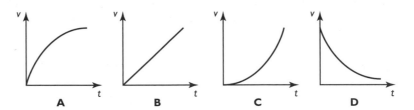

9 A rocket during the period just after take off? (1 mark)

10 A ball-bearing falling from rest through a viscous fluid? (1 mark)

Total: 10 marks

Answers to questions 1–10

(1) B

e Use $v = u + at$.

Convert the time to 100 s → $a = 0.06$ m s^{-2}

(2) C

e Use s = average velocity × time, or $v^2 = u^2 + 2as$ $s = 1300$ m

(3) C

e No direction can be assigned to energy and so it is a scalar quantity.

(4) D

e Horizontal component = 30 cos 30 m s^{-1} = 26 m s^{-1}

(5) C

e Strain is the ratio of extension (m) and original length (m) and so has no units.

(6) B

e Newton's first law states that a body will remain at rest or move with constant velocity unless there is a resultant force acting upon it. Equilibrium means that the resultant force is zero. It should be noted that answer D is not true unless the two forces are acting in opposite directions.

(7) A

e Hooke's law applies only to the linear region 0–P.

(8) B

e The elastic limit is usually just beyond the limit of proportionality, P, at point Q.

(9) C

e The thrust on the rocket will be constant over the early stage of the take-off, but the mass will be rapidly decreasing as fuel is burned. This leads to a bigger acceleration, represented by an increasing gradient on the velocity–time graph.

(10) A

e Initially the weight is bigger than the upthrust, so there will be a resultant downward force on the bearing causing it to accelerate. As the downward velocity increases so does the viscous drag ($F = 6\pi\eta rv$) and the acceleration reduces to zero when the sum of the upthrust and the drag equal the weight. The gradient of the graph will therefore continually reduce until it levels off.

Tip When analysing velocity–time graphs always look at how the gradient is changing — this represents the change in acceleration.

■ ■ ■

Question 11

The following is a passage from a report on the advantages and disadvantages of a stainless steel kettle compared with a plastic one. Complete the gaps in the passage by selecting appropriate words from the list below.

brittle	dense	elastic	harder	ions
plastic	polymers	stronger	tougher	

If the plastic kettle is dropped, or receives a sharp blow, it is more likely to crack or shatter than the steel one. This is because the material is (i) _____. The steel is much (ii) _____, and is capable of absorbing much more energy, and may only suffer from minor indentations caused by (iii) _____ deformation. The surface of the steel is (iv) _____ than the plastic and so less likely to be scratched.

The plastic kettle will retain its heat much better. It consists of long chain molecules, called (v) _____, which give it a low thermal conductivity. The plastic is also less (vi) _____ than the steel and so the kettle is lighter.

Total: 3 marks

■ ■ ■

Answers to question 11

(i) Brittle
(ii) Tougher
(iii) Plastic
(iv) Harder
(v) Polymers
(vi) Dense

e All six correct: 3 marks; 4 or 5 correct: 2 marks; 2 or 3 correct: 1 mark.

Question 12

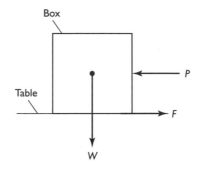

A box is pushed along a table at constant speed. The box is in *equilibrium*.

(i) **Explain the meaning of 'equilibrium'.** (1 mark)

(ii) **Copy the diagram and add the fourth force needed for the box to be in equilibrium.** (1 mark)

(iii) **State Newton's third law of motion.** (2 marks)

(iv) **Copy and complete the table.** (4 marks)

Force	Description of force	Body upon which the Newton's third law pair acts	Type of force	Direction of the Newton's third law pair force
W				
F				

Total: 8 marks

■ ■ ■

Answers to question 12

(i) Equilibrium is the condition of a body when the resultant force (in any two directions) is zero. ✓

(ii) Arrow drawn from the surface, upward and along the same line as the weight. ✓

(iii) If body A exerts a force on body B, then body B will exert an equal ✓ and opposite ✓ force on body A.

(iv)

Force	Description of force	Body upon which the Newton's third law pair acts	Type of force	Direction of the Newton's third law pair force
W	Weight	The Earth	Gravitational (NOT just gravity)	Up(ward)
F	Friction	Table	Contact	Right to left (opposite to F)

ℓ One mark for each correct column. ✓ ✓ ✓ ✓

■ ■ ■

Question 13

The graph shows the variation of velocity with time for a ball released from rest and allowed to bounce off the floor.

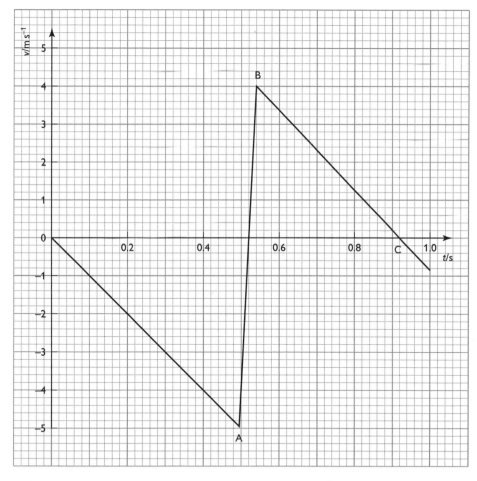

(i)   **Determine the gradient of the region 0A of the graph. What quantity does this represent?** (2 marks)

(ii) **Describe what is happening to the ball during the interval between A and B.** (1 mark)

(iii) **Calculate the displacement of the ball at point C.** (3 marks)

Total: 6 marks

■ ■ ■

Answers to question 13

(i) Gradient $= \dfrac{-5.0 \text{ m s}^{-1}}{0.50 \text{ s}} = -10 \text{ m s}^{-2}$ ✓

(*To gain this mark both the unit and the negative sign are needed.*)

The gradient represents acceleration (due to gravity). ✓

(ii) AB represents the time that the ball is in contact with the ground. ✓

(iii) Displacement is equal to the area under a velocity–time graph. ✓

$= \frac{1}{2}(0.50 \text{ s}) \times (-5.0 \text{ m s}^{-1}) + \frac{1}{2}(0.40 \text{ s}) \times (4.0 \text{ m s}^{-1})$ ✓ (*correct values regardless of sign*)

$= -1.25 \text{ m} + 0.80 \text{ m} = -0.45 \text{ m}$ ✓

e A C-grade candidate may gain some marks by finding a displacement using the area under the graph, but an A-grade candidate will recognise that displacement is a vector quantity. As downward velocities are assigned negative values, downward displacement and accelerations are also negative. In this case the displacement at point C is −0.45 m, i.e. 0.45 m below the point of release.

Question 14

A football is kicked off the ground towards the goal. The initial velocity of the ball has components of 20 m s^{-1} in the horizontal direction and 8.0 m s^{-1} in the vertical plane.

(a) Calculate:

    (i)   **the maximum height of the ball's trajectory**

    (ii) **the time taken to reach this height**   **(2 marks)**

(b) Show that the ball lands on the ground about 30 m from the kicker.   **(2 marks)**

(c) In practice, the ball travels less than 30 m before its first bounce. Suggest a reason for this.   **(1 mark)**

                                                                 Total: 5 marks

■ ■ ■

Answers to question 14

(a) (i) Use $v = u + at$

$0 = 8.0 \text{ m s}^{-1} - 9.8 \text{ m s}^{-2} \times t \rightarrow t = 0.82 \text{ s}$ ✓

    (ii) Use $v^2 = u^2 + 2as$

$0 = (8.0 \text{ m s}^{-1})^2 - 2 \times 9.8 \text{ m s}^{-2} \times s \rightarrow s = 3.3 \text{ m}$ ✓

e Note that the vertical component of the initial velocity is upward and the acceleration due to gravity always acts downward and so *g* is given a negative value in the equations.

(b) Horizontal displacement = horizontal component of velocity × time in the air

$x = 20 \text{ m s}^{-1} \times 1.64 \text{ s}$ ✓ $= 33 \text{ m}$ ✓

e This is a 'show that' question so the answer is shown to one more significant figure than that stated in the question.

(c) Range is less due to air resistance (*wind, spinning ball etc. acceptable*). ✓

■ ■ ■

Question 15

An airship (blimp) filled with helium has a total mass of 7000 kg and a volume of 6000 m³. It is moving slowly forward at a constant height through air with a density of 1.2 kg m⁻³.

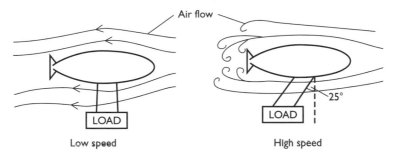

Low speed   High speed

(a) Calculate the mass of air displaced by the blimp, and hence determine the upthrust acting upon it. (2 marks)
(b) The airship is designed so that the air flowing over its surface moves in continuous layers. Give a word that describes this type of airflow. (1 mark)
(c) The airship accelerates to a speed where the air no longer moves uniformly across it, and the motion becomes bumpy. Give a name for this type of airflow. (1 mark)
(d) At the higher speed, the load supported by the two cables 'lags behind' the blimp, such that the cables make an angle of 25° to the vertical.
  (i) Draw a free-body force diagram for the airship. (3 marks)
  (ii) Calculate the tension in each of the cables. (2 marks)

Total: 9 marks

■ ■ ■

Answers to question 15

(a) mass = volume × density = 6000 m³ × 1.2 kg m⁻³ = 7200 kg ✓
   upthrust = weight of displaced air = 7200 kg × 9.8 m s⁻² = 71 000 N ✓

e It is a common error for candidates to calculate the upthrust as the *mass* of displaced fluid rather than the weight.

(b) Streamline or laminar flow ✓
(c) Turbulence, or turbulent flow ✓
(d) (i)

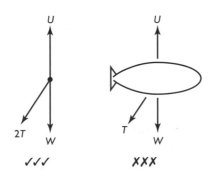

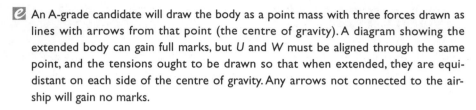 An A-grade candidate will draw the body as a point mass with three forces drawn as lines with arrows from that point (the centre of gravity). A diagram showing the extended body can gain full marks, but U and W must be aligned through the same point, and the tensions ought to be drawn so that when extended, they are equidistant on each side of the centre of gravity. Any arrows not connected to the airship will gain no marks.

(ii) The ship is in equilibrium in the vertical plane, hence:
Upthrust = weight of balloon (including helium) + vertical components of the tension
$71\ 000\ \text{N} = 7000\ \text{kg} \times 9.8\ \text{m s}^{-2} + 2 \times T \cos 25$ ✓
$T = 1100\ \text{N}$ ✓

■ ■ ■

Question 16

Fletcher's trolley was a common piece of equipment used in schools several decades ago. The trolley had five identical, cylindrical masses that were slotted into the body of the trolley. The masses were removed one at a time and added to the load, so that a range of forces could be applied to the trolley.

A horizontally vibrating steel strip was fitted to the strip so that a trace was drawn on a card as the trolley moved along the track.

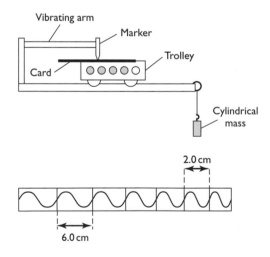

(a) Why is it necessary to add masses from the trolley to the load to increase the accelerating force? (1 mark)

(b) For the sample card shown, the frequency of the vibrations of the steel strip is 5.0 Hz — i.e. each complete oscillation takes a time of 0.20 s.

Use the distances given on the diagram to calculate the average velocity of the trolley during the first marked oscillation and during the fifth cycle. Hence determine the acceleration of the trolley. (3 marks)

(c) In schools nowadays, there are several variations of this experiment. Describe an alternative method of determining the acceleration of the trolley. List any additional apparatus that would be required, and what readings need to be taken. (3 marks)

(d) Explain how the acceleration may be calculated from the results. (2 marks)

(e) How can the results of the experiment be used to check the validity of Newton's second law of motion for bodies of fixed mass? (2 marks)

(f) In such an experiment the track is inclined slightly to compensate for friction. Why is this necessary for the law to be demonstrated? (2 marks)

Total: 13 marks

■ ■ ■

Answers to question 16

(a) The total mass that is being accelerated (trolley plus load) remains constant. ✓

(b)

$$\text{Initial velocity, } u = \frac{2.0 \times 10^{-2} \text{ m}}{0.20 \text{ s}} = 0.10 \text{ m s}^{-1} ✓$$

$$\text{Final velocity, } v = \frac{6.0 \times 10^{-2} \text{ m}}{0.20 \text{ s}} = 0.30 \text{ m s}^{-1} ✓$$

Time interval between u and v, $\Delta t = 4$ cycles $= 0.80$ s

$$\text{Acceleration } a = \frac{v - u}{\Delta t} = \frac{0.20 \text{ m s}^{-1}}{0.80 \text{ s}} = 0.25 \text{ m s}^{-2} ✓$$

(c) There are several possible methods. The four most likely are:

Ticker tape	Light gate/sensor	Motion sensor	Video/strobe
Ticker timer (tape and oscillator)	Timer/datalogger/PC	Datalogger/PC	Metre rule/markings on the track

A labelled diagram can gain both marks here. ✓ ✓

Description of the distance measured and any corresponding time or any mention of $v = \dfrac{d}{t}$ ✓

e Many candidates lose the final mark with incomplete statements like 'the time taken for the interrupter card to cut through the light beam is measured and so the velocity can be calculated'.

(d) Divide force by acceleration for all pairs of measurements, or plot a graph of force against acceleration. ✓

Ratio F/a is constant or the graph is a straight line through the origin. ✓

(e) Mention that two velocities need to be taken (or zero velocity at the start). ✓
Use of $v = u + at$ or $v^2 = u^2 + 2as$ to determine a. ✓

(f) Newton's second law relates to the resultant force acting on the trolley. ✓
For a level track, frictional forces will oppose the motion of the trolley, so that
the resultant force equals the load minus the friction. ✓

A C-grade candidate may state that 'friction will reduce the acceleration' and so only
gain the second mark.

It is a common complaint of examiners that the concept of resultant force is poorly
understood, and that candidates often state Newton's second law requires that
there are no external forces, or that inclining the track removes the external forces.

■ ■ ■

Question 17

A student uses a simple viscometer to investigate how the rate of flow of motor oil
through a glass tube depends on its diameter. She uses several pieces of tube of the same
length but with a range of internal diameters.
A diagram of the apparatus used by the student and the graph of her results is given below:

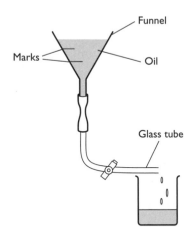
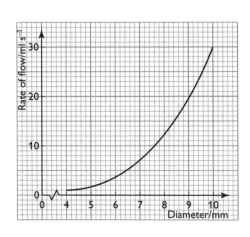

(a) Use the graph to find the rate of flow of oil for tubes of diameter 5 mm
and 10 mm. (2 marks)

(b) Suggest a possible relationship between the rate of oil flow and the internal
diameter of the tubes. (2 marks)

(c) State two other factors that will affect the rate of flow of the oil
through a glass tube. (2 marks)

Total: 6 marks

■ ■ ■

Answers to question 17

(a) When d = 5.0 mm the rate of flow = 2.0 ml s^{-1} and when d = 10 mm the rate of flow = 32 ml s^{-1}.

e One value ±½ of a scale division (2.0 ± 0.5 or 32 ± 0.5) ✓

e Both values correct including units. ✓

(b) The rate of flow through the larger diameter pipe is (about) 16 times that of the smaller one. ✓

This suggests that the flow rate is proportional to the fourth power of the diameter. ✓

e An A-grade candidate should be aware that if doubling the diameter increases the flow 16 times, a fourth power relationship is a possibility. Candidates who simply state that the flow rate increases will not gain any marks, but a C-grade candidate may gain 1 of the marks by referring to the values from the graph.

(c) Other factors that may affect the flow rate are:
- pressure (height of funnel)
- temperature (affects the viscosity)
- length of tube

e Any two. ✓ ✓

■ ■ ■

Question 18

A man pedals a bicycle along a level road with a constant velocity of 4.0 m s^{-1}. He stops pedalling and 'free-wheels' until the bike stops. The distance travelled while free-wheeling is 20 m, and the total mass of the man and the bike is 80 kg.

(a) Calculate:
(i) the average deceleration of the bike during free-wheeling (2 marks)
(ii) the average resistive force acting on the cycle (1 mark)
(b) Estimate the average power expended by the cyclist just before he stopped pedalling. State any assumptions you have made in calculating this value. (2 marks)

Total: 5 marks

■ ■ ■

Answers to question 18

(a) (i) $v^2 = u^2 + 2as$ $0 = (4.0 \text{ m s}^{-1})^2 + 2a \times 20 \text{ m}$
$a = (-) \, 0.40 \text{ m s}^{-2}$ ✓
(ii) $F = ma = 80 \text{ kg} \times 0.40 \text{ m s}^{-2} = 32 \text{ N}$ ✓
(b) $P = F \times v = 32 \text{ N} \times 4.0 \text{ m s}^{-2} = 130 \text{ W}$ ✓
It is assumed that the average resistive force is the same during pedalling as it is when free-wheeling to a halt. ✓

e A statement that the driving force equals resistive force when the bicycle is travelling at constant velocity is not strictly answering the question, but may be credited here.

■ ■ ■

Question 19

A bungee jumper dives off a bridge with an elastic cord tied to his ankles.

(a) Describe the energy changes that occur up to the time that the cord is fully extended. (3 marks)

(b) Explain why the jumper does not reach the level of the bridge when he is projected back upwards by the stretched cord. (2 marks)

Total: 5 marks

■ ■ ■

Answers to question 19

(a) During free-fall, the jumper loses gravitational potential energy and gains kinetic energy. ✓

e No marks will be awarded if just potential energy is given. The word 'gravitational' or an accepted acronym (GPE → KE) is essential.

During the period from when the cord becomes taut until the jumper momentarily stops when the cord is fully extended, the kinetic energy of the jumper AND a further drop in gravitational potential energy is transferred as elastic strain energy (or elastic potential energy) in the cord.

e ΔGPE + ΔKE ✓ (both needed) E_{el} (EPE) ✓

■ ■ ■

Question 20

A student carries out an investigation into the behaviour of copper wire, by clamping one end of the wire to the bench and applying varying loads to the other.

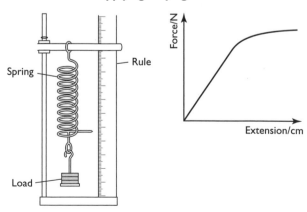

She sets up the equipment as shown in the diagram, with a 3.00 m length of wire, and measures the extensions of the wire for a range of loads up to 40.0 N.

(a) Why is it beneficial to use such a long piece of wire? (1 mark)

(b) The student intends to analyse her results using a stress–strain graph.

    (i) What additional measurement does she need to make?

    (ii) Define the terms stress and strain. (3 marks)

(c) A typical graph for such an experiment is drawn below. Use the graph to determine the Young modulus and the ultimate tensile stress (UTS) of copper. (3 marks)

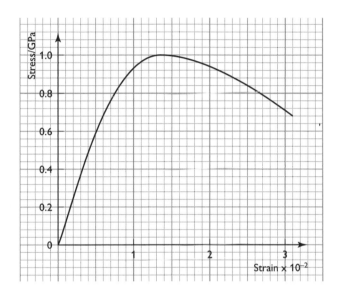

(d) As an extension to the investigation the experiment is repeated using a steel wire of the same dimensions as the original wire, and then using an equal length of thinner copper wire. Stress–strain graphs are drawn using the same scales as in the first investigation. Describe the appearance of these graphs compared with that of the initial experiment. (3 marks)

Total: 10 marks

Total for paper: 80 marks

■ ■ ■

Answers to question 20

(a)   Longer wire means bigger/more readable extension ✓ (for the same load).

(b) (i)   The diameter of the wire ✓

e A common error is to state that the cross-sectional area is measured. The area is required but is calculated from the diameter. Similarly, 'radius' would not gain the mark.

(ii)

$$\text{Stress} = \frac{\text{force}}{\text{cross-sectional area}} \checkmark \qquad \text{Strain} = \frac{\text{extension}}{\text{original length}} \checkmark$$

(c) Young modulus E = the gradient of the Hooke's law region

$$= \frac{0.6 \times 10^9 \text{ Pa}}{0.5 \times 10^{-3}} = 1.2 \checkmark \times 10^{11} \text{ Pa} \checkmark$$

e On an examination paper, where the answers are written on the paper, an A-grade candidate may extend the linear region of the graph so that a more accurate value of the gradient is obtained from the larger triangle.

UTS = 0.96 GPa $\checkmark$

(d) Stress–strain curve for steel will have a steeper linear region $\checkmark$ (because its Young modulus is bigger than copper) and will reach a much bigger stress $\checkmark$ (before it begins to yield).

The curve for the thinner copper wire will be identical to the first graph $\checkmark$ (stress–strain graphs show the properties of the material regardless of the dimensions of the sample, as these are incorporated in calculating stress and strain).

Test Paper 2

Time allowed 1 hour 20 minutes. Answer ALL the questions.

For questions 1–10 select one answer from A to D.

1 **Which of the following quantities is not a scalar quantity?**
   A displacement   **B** mass   **C** speed   **D** time   (1 mark)
2 **Which of the following does not affect the rate of flow of a fluid through a pipe when a constant pressure is applied across the ends of the pipe?**
   A density of the fluid
   B length of the pipe
   C radius of the pipe
   D temperature of the fluid   (1 mark)
3 **The pressure at the bottom of a column of liquid of height 80 cm and density 900 kg m^{-3} is about:**
   A 72 Pa   **B** 720 Pa   **C** 7.2 kPa   **D** 72 kPa   (1 mark)
4 **A high-jumper has an initial upward velocity of 6.0 m s^{-1}. The maximum height reached by her centre of gravity will be about:**
   A 1.6 m   **B** 1.8 m   **C** 2.0 m   **D** 2.2 m   (1 mark)

In questions 5 and 6, which of the following best completes the sentence?
A has an acceleration of zero
B has an acceleration of less than 9.8 m s^{-2}
C has an acceleration of 9.8 m s^{-2}
D has an acceleration of more than 9.8 m s^{-2}

Choose the appropriate letter that best completes the sentence. Each statement may be used once, more than once or not at all.

5 **A juggler's ball at the top of its trajectory ___ .**   (1 mark)
6 **A raindrop falling at its terminal velocity ___ .**   (1 mark)

In questions 7 and 8, which of the following graphs best represent the quantities described when they are plotted on the y- and x-axes? Each graph may be used once, more than once or not at all.

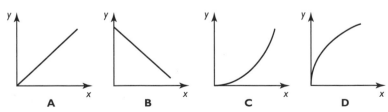

7 **y-axis: kinetic energy of a body accelerating from rest**
   x-axis: velocity of the body   (1 mark)
8 **y-axis: velocity of a car that is decelerating uniformly**
   x-axis: time after the application of the brakes   (1 mark)

In questions **9** and **10**, choose the appropriate line on the stress–strain graph that best describes the property of the material.

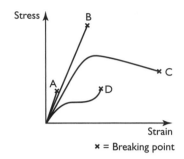

x = Breaking point

9 A material that is strong and brittle   (1 mark)
10 A material that is tougher than the others   (1 mark)

Total: 10 marks

■ ■ ■

Answers to questions 1–10

(1) A

 Displacement is the distance moved in a particular direction, and so is a vector quantity.

(2) A

Ⓔ The density has no effect. The temperature affects the viscosity, and hence the flow rate.

(3) C

Ⓔ $\Delta P = h\rho g = 0.80 \times 3900 \ \text{kg m}^{-3} \times 9.8 \ \text{m s}^{-2} = 7200 \ \text{Pa} = 7.2 \ \text{kPa}$

(4) B

Ⓔ $v^2 = u^2 + 2as \quad 0 = (6.0 \ \text{m s}^{-1})^2 - 2 \times 9.8 \ \text{m s}^{-2} \times s \quad s = 1.8 \ \text{m}$

(5) C

Ⓔ All objects in free motion on Earth accelerate downward at $9.8 \ \text{m s}^{-2}$, even though the velocity at the top is momentarily zero.

(6) A

Ⓔ The velocity is unchanging, so the acceleration is zero.

(7) C

Ⓔ Kinetic energy = $\frac{1}{2}mv^2$, so y is proportional to x^2.

(8) B

Ⓔ Acceleration is represented by the gradient. A uniform, negative gradient represents constant deceleration.

(9) B

e The line has the highest ultimate breaking stress and so is the strongest, and it has no plastic deformation so it is brittle.

(10) C

e There is a long region of plastic deformation and the area under the curve (energy absorbed) is the greatest, and so it represents the toughest material.

■ ■ ■

Question 11

An oarsman rows a boat across a river at right angles to the water flow at a velocity of 1.5 m s⁻¹. The water flows at 2.0 m s⁻¹.

(a) Draw a vector diagram to show the resultant velocity of the boat. (2 marks)

(b) By adding a scale to your diagram, or otherwise, determine the value of the resultant velocity. (2 marks)

Total: 4 marks

■ ■ ■

Answers to question 11

(a)

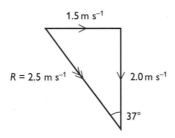

e Triangle correctly drawn 3 labelled with arrows correct ✓.

(b) $R^2 = (1.5 \text{ m s}^{-1})^2 + (2.0 \text{ m s}^{-1})^2$ $R = 2.5 \text{ m s}^{-1}$ ✓
Angle to the water flow = $\tan^{-1} 0.75 = 37°$ ✓

(Or both measured from a scale drawing).

e Velocity is a vector quantity and so it is essential to include the direction of the resultant.

■ ■ ■

Question 12

A teacher places a coin on the edge of a bench and another, identical, coin on the end of a metre rule, as shown in the diagram. The rule is moved sharply across the surface so that the coin on the rule is left behind and falls directly to the ground and the one on the bench is projected horizontally.

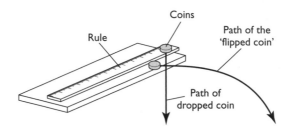

(a) Explain why both coins hit the floor at the same instant. (2 marks)

(b) The height of the bench is 0.82 m and the projected coin leaves the bench with a velocity of 3.2 m s^{-1}.

   (i) Show that the time taken for the coins to reach the floor is about 0.4 s. (3 marks)

   (ii) Calculate the horizontal distance travelled by the projected coin when it strikes the floor. (1 mark)

(c) The teacher repeats the demonstration but replaces the coin on the bench with a larger coin with a greater mass than the coin on the rule. Assuming the coin leaves the bench with a velocity of 3.2 m s^{-1}, as before, compare its trajectory with that of the coin projected in the first demonstration. Explain your answer. (3 marks)

Total: 9 marks

■ ■ ■

Answers to question 12

(a) Horizontal and vertical motions are independent of each other. ✓
Both coins accelerate downwards at the same rate (9.8 m s^{-2}). ✓

(b) (i) For the vertical motion use $s = ut + \frac{1}{2}at^2$ (or a combination of the two other equations). ✓
    0.82 m $= 0 + \frac{1}{2} \times 9.8$ m s$^{-2} \times t^2$ ✓ $\rightarrow t = 0.41$ s ✓

🅮 Note that this is a 'show that' question so the answer must be given to more than the one significant figure shown in the question.

   (ii) Horizontal displacement = velocity × time = 3.2 m s$^{-1} \times 0.41$ s = 1.3 m ✓

(c) The trajectory will be the same in both cases. ✓
The acceleration due to gravity is the same for all masses. ✓
(There is no component of the weight in the horizontal direction) so the horizontal velocity remains constant. ✓

■ ■ ■

Question 13

An electric motor is used to raise a load through a measured distance. A student uses the arrangement to investigate the power output of the motor. Her results are shown in the table:

Load/kg	Height/m	Time/s	Work done/J	Power/W
0.200	1.20	0.42		
0.400	1.20	0.78		

(a) Copy and complete the table to show the work done in raising the load and the output power for both loads. (2 marks)

(b) The motor is rated as 12 W. Assuming this value is the electrical input power, calculate the efficiency of the motor when raising the 0.400 kg load. (1 mark)

Total: 3 marks

■ ■ ■

Answers to question 13

(a) Work done = force distance = $0.200 \text{ kg} \times 9.8 \text{ m s}^{-2} \times 1.20 \text{ m} = 2.4 \text{ J}$
$= 0.400 \text{ kg} \times 9.8 \text{ m s}^{-2} \times 1.20 \text{ m} = 4.7 \text{ J}$ ✓

$$\text{Power} = \frac{\text{work done}}{\text{time}} = \frac{2.4 \text{ J}}{0.42 \text{ s}} = 5.6 \text{ W}$$

$$= \frac{4.7 \text{ J}}{0.78 \text{ s}} = 6.0 \text{ W} ✓$$

e Candidates who forget to include *g* in the calculation of work can still gain the second mark if the error is carried forward (gives 0.57 W and 0.61 W).

(b) $$\text{Efficiency} = \frac{\text{power output}}{\text{power input}} \times 100\% = \frac{6.0 \text{ W}}{12 \text{ W}} \times 100\% = 50\% ✓$$

■ ■ ■

Question 14

A man drags a rock of mass 60 kg up a slope of 20° at constant speed. A diagram of the forces acting on the rock is shown below.

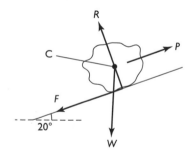

(a) **Name the point marked C on the diagram.** (1 mark)
(b) **State Newton's third law of motion.** (1 mark)
(c) **Give the Newton's third law pair force corresponding to (i) *W* and (ii) *F*.
In each case state the body on which it acts, the type of force and the
direction of the force.** (3 marks)

Force	Body upon which the Newton's third law pair acts	Type of force	Direction of the Newton's third law pair force
W			
F			

(d) **The weight, *W*, can be resolved into two components, one acting down the slope and
the other perpendicular to the slope.
Show that the component of the weight acting down the slope is about 200 N.** (1 mark)
(e) **The rock is said to be in equilibrium as it is being pulled up the slope.**
    (i) **Explain the meaning of equilibrium with reference to the moving rock.**
    (ii) **Calculate the magnitude of the frictional force, *F*, if the force, *P*, applied by
    the man to the rope is 250 N.** (4 marks)

Total: 10 marks

■ ■ ■

Answers to question 14

(a) C is the centre of gravity of the rock (*'centre of mass' is acceptable*). ✓
(b) Newton's third law states that if body A exerts a force on body B, then body B
    will exert an equal and opposite force on body A. ✓
(c)

Force	Body upon which the Newton's third law pair acts	Type of force	Direction of the Newton's third law pair force
W	The Earth	Gravitational	Up
F	The surface of the slope	Contract/ frictional	Up the slope/left to right

(3 marks)

e One mark for each correct column. ✓ ✓ ✓

(d) $60 \text{ kg} \times 9.8 \text{ m s}^{-2} \times \sin 20 = 201 \text{ N}$ ✓
(e)(i) The rock is in equilibrium because the resultant of all the forces acting on it
    is zero. ✓
    (ii) $P = F + W \sin 20$ ✓
        $F = 250 \text{ N} - 200 \text{ N}$ ✓ $= 50 \text{ N}$ ✓

■ ■ ■

Question 15

A student was asked the following question: 'Describe how the energy of a parachute jumper varies from the moment he leaves the aircraft until he reaches the ground.' As an answer, the student wrote the following: 'Initially the parachutist has potential energy which is converted into kinetic energy as he descends. When the parachute opens he continues to fall at a constant rate so his kinetic energy increases more slowly. Just before landing, his kinetic energy is equal to the initial potential energy, and all of this is lost when he hits the ground.'

(a) Discuss the student's answer, highlighting any incorrect or missing physics. (4 marks)

(b) The parachutist falls from a height of 1000 m to 980 m before the parachute opens. Calculate his velocity at the instant that the parachute opens. (2 marks)

Total: 6 marks

Answers to question 15

(a) There is no mention of 'gravitational' potential energy. ✓
When moving at a constant speed, the kinetic energy will remain the same. ✓
There is no mention that some of the gravitational potential energy is doing work against air resistance, or being transferred to the surrounding air during the descent. ✓
The final kinetic energy will not be equal to the initial gravitational potential energy. ✓
Energy is never 'lost'. It must be converted into other forms. ✓

e Any four of these five marking points will score maximum marks.

(b) Loss in GPE = gain in KE $mg\Delta h = \frac{1}{2}mv^2$ ✓
$v = \sqrt{(2g\Delta h)} = \sqrt{(2 \times 9.8 \text{ m s}^{-2} \times 20 \text{ m})} = 20 \text{ m s}^{-1}$ ✓

Question 16

(a) State Hooke's law. (1 mark)

(b) Make a list of the apparatus needed and the measurements taken for an investigation to test if a spring obeys Hooke's law. (2 marks)

(c) After such an experiment a student plots a graph that shows that the law is obeyed by the spring under test. Sketch a graph that would lead to this conclusion. Label both axes and draw the line representing the behaviour of the spring. (2 marks)

(d) **A further investigation is carried out using the same equipment, but with two identical springs connected firstly end to end (in series) to make a spring of twice the length, and secondly side by side (in parallel). On the same graph as that for the single spring draw two further lines, clearly labelling which combination is represented by each line.** (2 marks)

Total: 7 marks

■ ■ ■

Answers to question 16

(a) Hooke's law states that the extension is directly proportional to the applied load ✓ (provided a permanently deforming load is not applied).

(b) Equipment: slotted masses/weights, ruler/scale, supporting apparatus (clamp, stand etc.) ✓

Measurements: load/weight, extension ✓

(c)

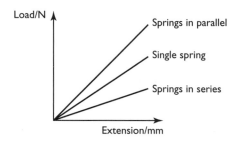

🅔 Both axes correctly labelled ✓ straight line through the origin ✓.

(d) *Straight lines drawn above and below the original line correctly labelled.* ✓
Gradients approximately double and one half of the original line. ✓

🅔 An A-grade candidate will be aware that the spring constant doubles when two identical springs (or wires) are joined in parallel, and halve when they are connected end to end. Simply showing that the stiffness increases or decreases is insufficient for full marks to be gained.

■ ■ ■

Question 17

A car of mass 1500 kg tows a trailer of mass 1000 kg along a level road. The driving force on the car is 2000 N and the total resistive forces on the car and trailer are 500 N and 300 N respectively.

(a) Calculate the resultant force acting on the combination. (1 mark)

(b) Determine the acceleration of the car and trailer. (3 mark)

(c) How long will it take the car and trailer to accelerate from 10 m s^{-1} to 20 m s^{-1} if the forces remain the same during this period? (2 marks)

(d) Draw a diagram of the trailer showing the forces acting upon it in the horizontal direction. (2 marks)

(e) Calculate the force the car exerts on the trailer through the coupling. (3 marks)

Total: 11 marks

■ ■ ■

Answers to question 17

(a) Resultant force = 2000 N − (500 N + 300 N) = 1200 N ✓

(b) By Newton's second law, for a fixed mass, $\Sigma F = ma$ ✓
1200 N = (1500 kg + 1000 kg) × a ✓
a = 0.80 m s^{-2} ✓

(c) Use $v = u + at$ 20 m s^{-1} = 10 m s^{-1} + 0.80 m s^{-2} × t ✓
t = 13 s ✓

(d)

Pull of car   ◄─ ┄┄┄┄ ─►   Resistive forces
P   F

Pull of car, P ✓ Resistive forces, F ✓

(e) Resultant force = $P − F = P − 300$ N ✓ = 1000 kg × 0.80 m s^{-2} ✓
P = 1100 N ✓

e Many candidates lose marks on Newton's second law questions by failing to calculate the resultant force, and just including the driving force in the equation $F = ma$.

■ ■ ■

Question 18

A rubber band is loaded by adding weights one at a time, and the extension of the band is measured every time the load is increased. The results are shown on the force–extension graph below:

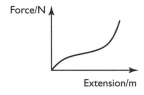

Force/N

Extension/m

(a) A rubber band is often called an *elastic* band. In what way does the band behave to suggest that this is an apt description? (1 mark)

(b) For large loads the band becomes quite *stiff*. Use the graph to explain the meaning of stiffness. (2 marks)

(c) **When the weights are removed from the band the rubber contracts. Copy the graph and draw a second line to represent the unloading curve. Label both curves.**

(1 mark)

(d) (i) **How could you estimate the elastic strain energy stored in the loaded spring?**

(1 mark)

(ii) **What does the unloading curve tell you about the work done on the weights when the band is unloaded, compared with the elastic strain energy at maximum extension? Explain your answer.**

(2 marks)

Total: 7 marks

Answers to question 18

(a) It is elastic because it will return to its original length when it is unloaded. ✓

(b) A stiff material requires a large force to produce a small extension. ✓
The stiff region on the graph has a relatively high gradient. ✓

e When a question asks you to 'use the graph', you will lose the second mark if you do not refer to the graph.

(c)

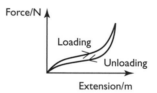

Both lines drawn and correctly labelled. ✓

(d) (i) The elastic strain energy is represented by the area under the curve. ✓

(ii) The work done by the band is less than the total elastic strain energy. ✓
Some energy is converted into other forms (e.g. heat/internal energy in the rubber). ✓

Question 19

The diagrams show the cross-section of the wing of a model aeroplane in level flight and while ascending. The lines represent the motion of layers of air relative to the wing.

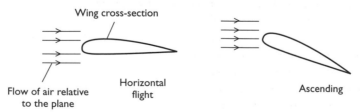

(a) In level flight the flow of air across the wings is streamlined (or laminar), but if the angle of ascent is too great, turbulence may occur and the aeroplane may stall. **Copy the diagram and complete the airflow lines above and below both wings.** (2 marks)

(b) The model aeroplane has a mass of 250 g and the surface area of the underside of its wings is 0.10 m². **Show that the pressure difference between the underside of the wing and upper surface is about 25 Pa when the model is flying horizontally.** (3 marks)

Total: 5 marks

■ ■ ■

Answers to question 19

(a)

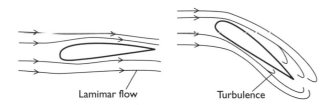

Lamimar flow   Turbulence

Laminar flow: continuous lines above and below the wing. ✓
Turbulent flow: some 'swirling' or break in continuity of the lines. ✓

(b)
$$\text{Pressure difference, } \Delta P = \frac{\text{upward force on wings (uplift)}}{\text{surface area of underside of wings}}$$

In level flight, the uplift = the weight of the aeroplane = 0.250 kg × 9.8 m s⁻² ✓

$$\Delta P = \frac{0.250 \text{ kg} \times 9.8 \text{ m s}^{-2}}{0.10 \text{ m}^2} \checkmark = 24.5 \text{ Pa} \checkmark$$

e This is a 'show that' question and so the answer is given to three significant figures. Candidates often lose the first mark by using the mass of the aeroplane instead of the weight.

■ ■ ■

Question 20

(a) **Draw a free-body force diagram of a sphere falling through a fluid at its terminal velocity. Label your diagram naming all the forces acting on the sphere.** (3 marks)

(b) **With reference to the diagram, explain the meaning of terminal velocity.** (2 marks)

(c) **A student carried out an investigation into the variation of the viscosity of syrup with its temperature. The terminal velocity of a 4.0 mm diameter ball bearing was found by timing the ball as it moved between two marks drawn 10.0 cm apart on the side of a test tube. This was repeated for a range of temperatures and the corresponding values of the viscosity of the syrup were calculated.**

In her conclusion the student stated 'The accuracy of the experiment, particularly at the higher temperatures, could be improved by using 2 mm diameter balls and a longer test tube.'

Comment on this conclusion, explaining if, and how, the accuracy of the terminal velocity measurements would be affected by these changes. **(3 marks)**

Total: 8 marks

Total for paper: 80 marks

■ ■ ■

Answers to question 20

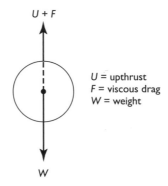

U + F

U = upthrust
F = viscous drag
W = weight

W

(a) *Weight, upthrust and viscous drag (resistive force or simply drag OK) correctly drawn and labelled.* ✓ ✓ ✓

(b) At terminal velocity the resultant force on the sphere is zero. ✓
On the diagram: $W = U + F$; or the upward forces on the diagram equal the downward forces etc. ✓

ℓ To gain the second mark, some reference must be made to the free-body force diagram.

(c) If the terminal velocity is high, the time taken for the ball to fall between the marks will be small, leading to a greater uncertainty (error) in the value of the terminal velocity, and hence in the viscosity. ✓
The time will be smallest at the higher temperatures because the viscosity will be lower and so the terminal velocity will be bigger. ✓
The time can be made bigger by using smaller balls or by increasing the length between the timing marks (*either* ✓).

ℓ An A-grade candidate would answer this question by referring to the percentage uncertainties in the readings. Such a candidate may also realise that the terminal velocity depends on the square of the radius, and state that a sphere of half the radius will travel four times more slowly.